Distributed and Parallel Architectures for Spatial Data

Distributed and Parallel Architectures for Spatial Data

Editors

Alberto Belussi
Sara Migliorini
Damiano Carra
Eliseo Clementini

MDPI • Basel • Beijing • Wuhan • Barcelona • Belgrade • Manchester • Tokyo • Cluj • Tianjin

Editors
Alberto Belussi
University of Verona
Italy

Sara Migliorini
University of Verona
Italy

Damiano Carra
University of Verona
Italy

Eliseo Clementini
University of L'Aquila
Italy

Editorial Office
MDPI
St. Alban-Anlage 66
4052 Basel, Switzerland

This is a reprint of articles from the Special Issue published online in the open access journal *ISPRS International Journal of Geo-Information* (ISSN 2220-9964) (available at: https://www.mdpi.com/journal/ijgi/special_issues/distributed_parallel_architectures_spatial_data).

For citation purposes, cite each article independently as indicated on the article page online and as indicated below:

LastName, A.A.; LastName, B.B.; LastName, C.C. Article Title. *Journal Name* **Year**, *Article Number*, Page Range.

ISBN 978-3-03936-750-4 (Hbk)
ISBN 978-3-03936-751-1 (PDF)

Contents

About the Editors

Alberto Belussi received his master's degree in Electronic Engineering in 1992 and PhD degree in Computer Engineering in 1996, both from the Politecnico di Milano. Since 1998, he has been based at University of Verona (Italy), where he has served as Associate Professor since his appointment in 2004. His main research interests include conceptual modeling of spatial databases, geographical information systems, spatial data integration, and spatial query optimization.

Sara Migliorini received her master's and PhD degree in Computer Science in 2007 and 2012, respectively, both from the University of Verona. Since 2012, she has been working at University of Verona (Italy), where she was appointed as Postdoctoral Research Associate from 2012 to 2019 and is currently serving as Assistant Professor. Her main research interests include geographic information systems, scientific workflow systems, and collaborative and distributed architectures.

Damiano Carra received his Laurea in Telecommunication Engineering from Politecnico di Milano, and his PhD in Computer Science from University of Trento. He is currently an Associate Professor in the Computer Science Department at University of Verona. His research interests include modeling and performance evaluation of large-scale distributed systems.

Eliseo Clementini is Associate Professor of Computer Science at the Department of Industrial and Information Engineering and Economics of the University of L'Aquila (Italy). He received his M.Eng. in Electronics Engineering from University of L'Aquila in 1990 and his PhD in Computer Science from University of Lyon in 2009. His research interests are mainly in the fields of spatial databases and geographical information science.

Preface to "Distributed and Parallel Architectures for Spatial Data"

In recent years, an increasing amount of spatial data has been collected by different types of devices, such as mobile phones, sensors, satellites, space telescope, and medical tools for analysis, or is generated by social networks, such as geotagged tweets. The processing of this huge amount of information, including spatial properties, which are frequently represented in heterogeneous ways, is a challenging task that has boosted research in the big data area in an attempt to investigate cases and propose new solutions for dealing with its peculiarities.

In the literature, many different proposals and approaches for facing the problem have been proposed, addressing different goals and different types of users. However, most are obtained by customizing existing approaches which were originally developed for the processing of big data of the alphanumeric type, without any specific support for spatial or spatiotemporal properties. Thus, the proposed solutions can exploit the parallelism provided by these kinds of systems, but without taking into account, in a proficient way, the space and time dimensions that intrinsically characterize the analyzed datasets. As described in the literature, current solutions include: (i) the on-top approach, where an underlying system for traditional big datasets is used as a black box while spatial processing is added through the definition of user-defined functions that are specified on top of the underlying system; (ii) the from-scratch approach, where a completely new system is implemented for a specific application context; and (iii) the built-in approach, where an existing solution is extended by injecting spatial data functions into its core.

This book aims at promoting new and innovative studies, proposing new architectures or innovative evolutions of existing ones, and illustrating experiments on current technologies in order to improve the efficiency and effectiveness of distributed and cluster systems when they deal with spatiotemporal data.

Alberto Belussi, Sara Migliorini, Damiano Carra, Eliseo Clementini
Editors

International Journal of **Geo-Information**

Article

Distributed Processing of Location-Based Aggregate Queries Using MapReduce

Yuan-Ko Huang

Department of Maritime Information and Technology, National Kaohsiung University of Science and Technology, 80543 Kaohsiung City, Taiwan; huangyk@nkust.edu.tw

Received: 17 July 2019; Accepted: 19 August 2019; Published: 23 August 2019

Abstract: The *location-based aggregate queries*, consisting of the *shortest average distance query (SAvgDQ)*, the *shortest minimal distance query (SMinDQ)*, the *shortest maximal distance query (SMaxDQ)*, and the *shortest sum distance query (SSumDQ)* are new types of location-based queries. Such queries can be used to provide the user with useful object information by considering both the spatial closeness of objects to the query object and the neighboring relationship between objects. Due to a large amount of location-based aggregate queries that need to be evaluated concurrently, the centralized processing system would suffer a heavy query load, leading eventually to poor performance. As a result, in this paper, we focus on developing the distributed processing technique to answer multiple location-based aggregate queries, based on the *MapReduce* platform. We first design a grid structure to manage information of objects by taking into account the storage balance, and then develop a distributed processing algorithm, namely the *MapReduce-based aggregate query algorithm (MRAggQ algorithm)*, to efficiently process the location-based aggregate queries in a distributed manner. Extensive experiments using synthetic and real datasets are conducted to demonstrate the scalability and the efficiency of the proposed processing algorithm.

Keywords: location-based aggregate queries; distributed processing technique; MapReduce; grid structure; MapReduce-based aggregate query algorithm

---

1. Introduction

With the fast advances of ubiquitous and mobile computing, processing the location-based queries on spatial objects [1–6] has become essential for various applications, such as traffic control systems, location-aware advertisements, and mobile information systems. Currently, most of the conventional location-based queries focus exclusively on a single type of objects (e.g., the nearest neighbor query finds a closest restaurant or hotel to the user). In other words, the different types of objects (termed *the heterogeneous objects*) are independently considered in processing the location-based queries, which means that the neighboring relationship between the heterogeneous objects is completely ignored. Let us consider a scenario where the user wants to stay in a hotel, have lunch in a restaurant, and go to the movies. Here, the hotel, the restaurant, and the theater refer to the heterogeneous objects. If the nearest neighbor queries are independently processed for the heterogeneous objects, the user is able to know his/her closest hotel, restaurant, and theater, which, however, may actually be far away from each other. Therefore, in addition to the spatial closeness of the heterogeneous objects to the query point, the neighboring relationship between the heterogeneous objects should also play an important role in determining the query result.

In the previous work [7], we present the *location-based aggregate queries* to provide information of the heterogeneous objects by taking into account both the neighboring relationship and the spatial closeness of the heterogeneous objects. In order to preserve the neighboring relationship between the heterogeneous objects, the location-based aggregate queries aim at finding the heterogeneous objects closer to each other by constraining their distance to be within a user-defined distance d. The set of objects satisfying the constraint of distance d is termed *the heterogeneous neighboring object set* (or *HNO set*). On the other hand, for maintaining the spatial closeness of the heterogeneous objects to the query point, four types of location-based aggregate queries are presented to provide information of *HNO set* according to specific user requirement. They are *the shortest average-distance query* (or *SAvgDQ*), *the shortest minimal-distance query* (or *SMinDQ*), *the shortest maximal-distance query* (or *SMaxDQ*), and *the shortest sum-distance query* (or *SSumDQ*), which are described respectively as follows.

- Consider the n types of objects, O_1, O_2, ..., O_n. Assume that there are m HNO sets, $\{o_1^1, o_2^1, ..., o_n^1\}$, $\{o_1^2, o_2^2, ..., o_n^2\}$, ..., $\{o_1^m, o_2^m, ..., o_n^m\}$, where $o_i^j \in O_i$, $i = 1 \sim n$, and $j = 1 \sim m$. Given a query point q, a set of objects $\{o_1^j, o_2^j, ..., o_n^j\}$ among these m HNO sets is determined, such that

  - for the *SAvgDQ*, the average distance of $\{o_1^j, o_2^j, ..., o_n^j\}$ to q is equal to

$$\min\{\frac{1}{n}(\sum_{i=1}^{n} d(q, o_i^j)) \mid j = 1 \sim m\},$$

  where $d(q, o_i^j)$ refers to the distance between objects o_i^j and q.
  - for the *SMinDQ*, the distance of an object $o_i^j \in \{o_1^j, o_2^j, ..., o_n^j\}$ to q is equal to

$$\min\{\min\{d(q, o_i^j) \mid i = 1 \sim n\} \mid j = 1 \sim m\}.$$

  - for the *SMaxDQ*, the distance of an object $o_i^j \in \{o_1^j, o_2^j, ..., o_n^j\}$ to q is equal to

$$\min\{\max\{d(q, o_i^j) \mid i = 1 \sim n\} \mid j = 1 \sim m\}.$$

  - for the *SSumDQ*, the traveling distance from q to $\{o_1^j, o_2^j, ..., o_n^j\}$ is equal to

$$\min\{d(q, \{o_1^j, o_2^j, ..., o_n^j\}) \mid j = 1 \sim m\},$$

  where $d(q, \{o_1^j, o_2^j, ..., o_n^j\})$ is the shortest distance that, starting from q, visits each object in $\{o_1^j, o_2^j, ..., o_n^j\}$ exactly once.

Let us use Figure 1 to illustrate how to process the four types of location-based aggregate queries (i.e., the *SAvgDQ*, the *SMinDQ*, the *SMaxDQ*, and the *SSumDQ*). As shown in Figure 1a, there are three types of data objects in the space, the hotels h_1 to h_5, the restaurants r_1 to r_5, and the theaters t_1 to t_5. Assume that the user-defined distance d is set to 2 (that is, the distance between any two objects should be less than or equal to 2), which leads to three *HNO sets*, $\{h_1, r_3, t_1\}$, $\{h_2, r_1, t_3\}$, and $\{h_3, r_2, t_2\}$ (shown as the gray areas). Take the query point q_1 in Figure 1b, issuing the *SAvgDQ*, as an example. For each *HNO set*, the distance between each object in the *HNO set* and the query point q_1 needs to be first computed and then the *HNO set* with the shortest average-distance to q_1 is the result set of the *SAvgDQ* (i.e., the set $\{h_2, r_1, t_3\}$). Meanwhile, the *SMinDQ* and the *SMaxDQ* issued by the query points q_2 and q_3, respectively, also need to be evaluated. When the *SMinDQ* is considered, the distances of the objects closest to q_2 in $\{h_1, r_3, t_1\}$, $\{h_2, r_1, t_3\}$, and $\{h_3, r_2, t_2\}$, respectively, are compared to each other, and then the *HNO set* (i.e., $\{h_3, r_2, t_2\}$) containing q_2's nearest neighbor is returned as the result set. In contrast to the *SMinDQ*, the *SMaxDQ* takes the furthest object in each *HNO set* into account. For the query point q_3, its furthest objects in the three *HNO sets* are t_1, t_3, and t_2, respectively. Among them, object t_1 has the shortest distance to q_3, and hence the *SMaxDQ* retrieves the set $\{h_1, r_3, t_1\}$ because it contains t_1. Consider the *SSumDQ* issued from the query point q_4, which is processed simultaneously by the system. The shortest traveling path for each of the three *HNO sets* $\{h_1, r_3, t_1\}$, $\{h_2, r_1, t_3\}$, and $\{h_3, r_2, t_2\}$ has to be determined so as to find the *HNO set* resulting in a shortest traveling distance from q_4. Finally, the set $\{h_1, r_3, t_1\}$ can be the *SSumDQ* result because of its shortest path $q_4 \to h_1 \to r_3 \to t_1$.

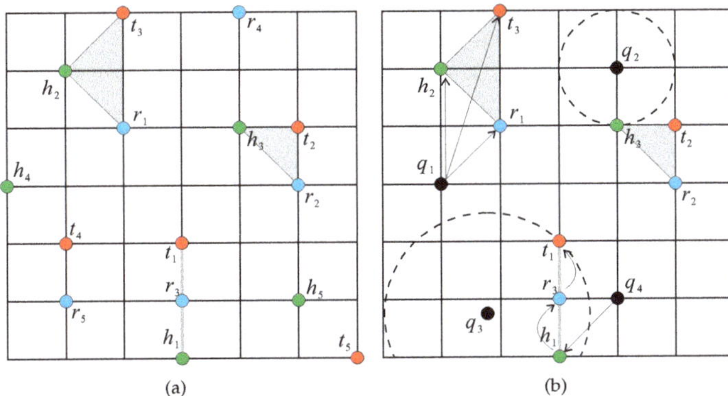

Figure 1. Example of processing the location-based aggregate queries. (**a**) Heterogeneous objects; (**b**) Multiple queries.

The processing techniques developed in [7] focus only on efficiently processing a location-based aggregate query (corresponding to *SAvgDQ*, *SMinDQ*, *SMaxDQ*, or *SSumDQ*). However, in highly dynamic environments, where users can obtain object information through the portable computers (e.g., laptops, 3G mobile phones, and tablet PCs), multiple location-based aggregate queries must be issued by the users from anywhere and anytime (For instance, in Figure 1, the *SAvgDQ*, the *SMinDQ*, the *SMaxDQ*, and the *SSumDQ* are issued from different query points at the same time.) It means that, when there is a large number of location-based aggregate queries processed concurrently, the time spent on sequentially evaluating the location-based aggregate queries would dramatically increase. Even worse, at the time at which a location-based aggregate query terminates, the query result may already be outdated. As a result, it is necessary to design the distributed processing techniques to rapidly evaluate multiple location-based aggregate queries.

To achieve the objective of distributed processing of location-based aggregate queries, we adopt the most notable platform, *MapReduce* [8], for processing multiple queries over large-scale datasets by involving a number of share-nothing machines. For data storage, an existing distributed file system (DFS), such as Google File System (GFS) or Hadoop Distributed File System (HDFS), is usually used as the underlying storage system. Based on the partitioning strategy used in the DFS, data are divided into equal-sized chunks, which are distributed over the machines. For query processing, the MapReduce-based algorithm executes in several *jobs*, each of which has three phases: *map*, *shuffle*, and *reduce*. In the map phase, each participating machine prepares information to be delivered to other machines. As for the shuffle phase, it is responsible for the actual data transfer. In the reduce phase, each machine performs calculation using its local storage. The current job finishes after the reduce phase. If the process has not been completed, another MapReduce job starts. Depending on the applications, the MapReduce job may be executed once or multiple times.

In this paper, we focus on developing the MapReduce-based methods to efficiently answer multiple location-based aggregate queries (consisting of numerous *SAvgDQ*, *SMinDQ*, *SMaxDQ*, and *SSumDQ* issued concurrently from different query points) in a distributed manner. We first utilize a grid structure to manage the heterogeneous objects in the space by taking into account the storage balance, and information of the partitioned object data in each grid cell is stored in the DFS. Next, we propose a distributed processing algorithm, namely the *MapReduce-based aggregate query algorithm* (*MRAggQ algorithm* for short), which is composed of four phases: the *Inner HNO set determining phase*, the *Outer HNO set determining phase*, the *Aggregate-distance computing phase*, and the *Result set generating phase*, each of which executes a MapReduce job to finish the procedure. Finally, we conduct a comprehensive set of experiments over synthetic and real datasets, demonstrating the efficiency, the robustness, and the scalability of the proposed *MRAggQ* algorithm, in terms of the average running time in performing different workloads of location-based aggregate queries.

The rest of this paper is organized as follows. In Section 2, we review the previous work on processing various types of location-based queries in centralized and distributed environments. Section 3 describes the grid structure used for maintaining information of the heterogeneous objects. In Section 4, we present how the *MRAggQ* algorithm can be used to process multiple location-based aggregate queries efficiently. Section 5 shows extensive experiments on the performance of the proposed methods. In Section 6, we conclude the paper with directions on future work.

2. Related Works

Efficient processing of the location-based queries is an emerging research topic in recent years. Here, we first review the centralized methods for processing the location-based queries on a single object type and multiple types of objects (i.e., the heterogeneous objects). Then, we discuss the MapReduce programming technique and survey some works on processing the location-based queries using MapReduce.

2.1. Centralized Processing Techniques for Location-Based Queries

Most of the conventional location-based queries on a single data type concentrate on discovering the spatial closeness of objects to the query object. The range query [9,10] is a well-known query, used to find a set of objects that are inside a spatial region specified by the user. If the spatial region is constructed according to the location of the query object q, another variation of range query, the within query [11,12], is presented to find the objects whose distances to q are less than or equal to a user-given distance d (i.e., finding the objects within the region centered at q with radius d). Recently, many efforts have been made on processing the range and within queries in different research domains, such as mobile information systems [3,13] and uncertain database systems [2,14]. The nearest neighbor query [15,16] is the most common type of location-based queries, as it has important applications to the provision of location-based services. Many variations of nearest neighbor query have been proposed in numerous applications. To address the issue of scalability, the *KNN* join query [17,18] is presented to find the K-nearest neighbors for all objects in a query set. To express requests by groups of users, the aggregate nearest neighbor (ANN) query (a.k.a. group nearest neighbor query) is proposed by Papadias et al. [19]. Given a set of query objects Q and a set of objects O, ANN query returns the object in O minimizing an aggregate distance function (e.g., sum or max) with respect to the objects in Q. A variation of nearest neighbor query with asymmetric property is the reverse nearest neighbor (RNN) query [1]. Given the query object q, the RNN query retrieves the set of objects whose nearest neighbor is q. The skyline query, also known as the maximal vector problem [20,21], is first studied in the area of computational geometry. Then, Borzsonyi et al. [22] introduce the skyline operator into database systems. If an object is not dominated by any other objects in terms of multiple attributes, then it is a *skyline point*. By taking into account the object locations, the spatial skyline query [4] is proposed, where the distance of objects plays an important role in determining the skyline points. Given a set of m query objects and a set of n data objects, each data object has m attributes, each of which refers to its distance to a query object. The spatial skyline query retrieves the skyline points that are not dominated in terms of the m attributes.

Some related work on processing the location-based queries tries to keep the neighboring relationship between the heterogeneous objects. Given two types of data objects A and B, the K closest pair query [23] finds the K closest object pairs between A and B (that is, the K pairs (a, b), where $a \in A$ and $b \in B$, with the smallest distance between them). Another type of location-based queries on the two data sources is the spatial join query [24], which maintains a set of object pairs (each pair has one item from the two data sources respectively) satisfying a given spatial predicate (e.g., *overlap* or *coverage*). Papadias et al. [25] further extend the spatial join query to the multiway spatial join query, in which the spatial predicate is a function over m data sources (where $m \geq 2$). Zhang et al. [26] present the *KNG* query to determine the query result based on (1) the *minimum* distance between the heterogeneous objects and the query object (referred to as *inter-group distance*) and (2) the *maximum* distance among the heterogeneous objects (referred to as *inner-group distance*). Given a spatial database with m types of data objects and a query object q, the KNG query returns the K groups (each of which consists of one object from each data type) with the minimum sum of the inner-group distance and the inter-group distance. However, due to the fact that the KNG query considers the sum of inner-group and inter-group distances, the object group retrieved by executing the KNG query is likely to be close to the query object but far away from each other (i.e., the inter-group distance dominates the query result), or close to each other but far away from the query object (i.e., the inner-group distance affects the result). To appropriately keep the spatial closeness and the neighboring relationship of objects, in our previous work [7], the location-based aggregate queries are presented to obtain information of the NHO sets.

2.2. Distributed Processing Techniques for Location-Based Queries

As mentioned in Section 1, MapReduce is a popular programming framework, which can be used to support the distributed processing of location-based queries. A MapReduce algorithm proceeds in several jobs, each of which has the map, the shuffle, and the reduce phases. In the map phase, for each participating machine, a list of key-value pairs (k, v) is generated from its local storage, where the key k is usually numeric and the value v corresponds to arbitrary information. According to the key k, each pair (k, v) is transmitted to another machine in the shuffle phase. More specifically, the shuffle phase distributes the key-value pairs across the machines following the rule that pairs with the same key are delivered to the same machine. In the reduce phase, each machine incorporates the key-value pairs received form the shuffle phase into its local storage, and performs the task using the local data. When the reduce phases of all machines are completed, the current MapReduce job terminates.

There has been considerable interest on supporting location-based queries over MapReduce framework. Cary et al. [27] present the techniques for building R-trees based on MapReduce, which, however, do not address the issues of processing the location-based queries. Zhang et al. [28] show how the location-based queries can be naturally expressed in MapReduce framework, including the spatial selection queries, the spatial join queries, and the nearest neighbor queries. Ji et al. [29] propose a MapReduce-based approach, in which an inverted grid structure is built to index data objects, to answer the KNN queries. Furthermore, in [30], they extend their approach to process a variant of KNN queries, the RKNN query. Akdogan et al. [31] focus on processing various types of location-based queries (including RNN, MaxRNN, and KNN queries), by creating a Voronoi diagram based on the MapReduce programming model for data objects. In their method, each data object is represented as a pivot which is then used to partition the space. Yokoyama et al. [32] propose a method that decomposes the given space into cells and evaluates the AKNN queries using MapReduce in a distributed and parallel manner. Zhang et al. [33] present the exact and approximate MapReduce-based algorithms to efficiently perform parallel KNN join queries on a large-scale dataset. To improve the performance of KNN join queries, Lu et al. [34] further design an effective mapping mechanism, by exploiting pruning rules for distance filtering, to reduce both the shuffling and computational costs.

Recently, Eldawy et al. [35,36] focus on developing a MapReduce framework, the *SpatialHadoop*, which is a comprehensive extension of Hadoop. The SpatialHadoop provides an expressive high level language for spatial objects, adapts a set of spatial index structures (e.g., Grid structure, R-tree, and R^+-tree) which is built-in HDFS, and supports the traditional location-based queries (including the range, KNN, and spatial join queries). Moreover, in [37], they address the issue of processing the skewed distributed datasets in the SpatialHadoop, by presenting a box counting function to detect the degree of skewness of a spatial dataset. The SpatialHadoop is carefully designed for the location-based queries, in which the spatial closeness of a single type of objects to the query point is a main concern in determining the query result. However, it cannot directly be applied for answering the location-based aggregate queries because (1) the query result consists of the heterogeneous objects, rather than a single type of objects, and (2) whether the heterogeneous objects satisfy the constraint of distance d (i.e., with the better neighboring relationship) should be taken into account.

3. Grid Structure

In our model, there are n types of data objects (i.e., the heterogeneous objects) in the space. As the location database contains large amounts of information that need to be maintained, a grid structure is used to manage such information by partitioning the space into multiple gird cells, each of which stores data of objects enclosed in it. In order to balance the storage load of each grid cell, the data space is partitioned into $C \times C$ equal-sized cells by considering a pre-defined parameter α. Initially, all the heterogeneous objects are grouped into 1×1 cells. Then, the number of objects enclosed in a cell is compared with the parameter α. Once the object number is greater than α, the data space covering all objects is repartitioned into 2×2 cells. Similarly, if there still exists a cell within which the object number exceeds α, then the data space needs to be repartitioned into 3×3 cells. This partitioning process continues until each cell $cell(c)$ satisfies the condition that the number of objects in $cell(c)$ is less than or equal to α. By exploiting the parameter α, the storage overhead for maintaining information of objects can be evenly distributed among the cells. Figure 2 shows an example of how the data space is divided by taking into account the storage load of each cell. As shown in Figure 2a, there are three types of data objects, R, S, and T in the space, each of which has five objects with coordinate (x, y) (e.g., object r_1's coordinate (x, y) refers to $(3, 14)$). Suppose that the pre-defined parameter α is set to 3. The data space would be divided into 3×3 cells, so as to guarantee that the number of objects in each cell does not exceed 3. The final divided grid cells, which are numbered from 0 to 8, are shown in Figure 2b.

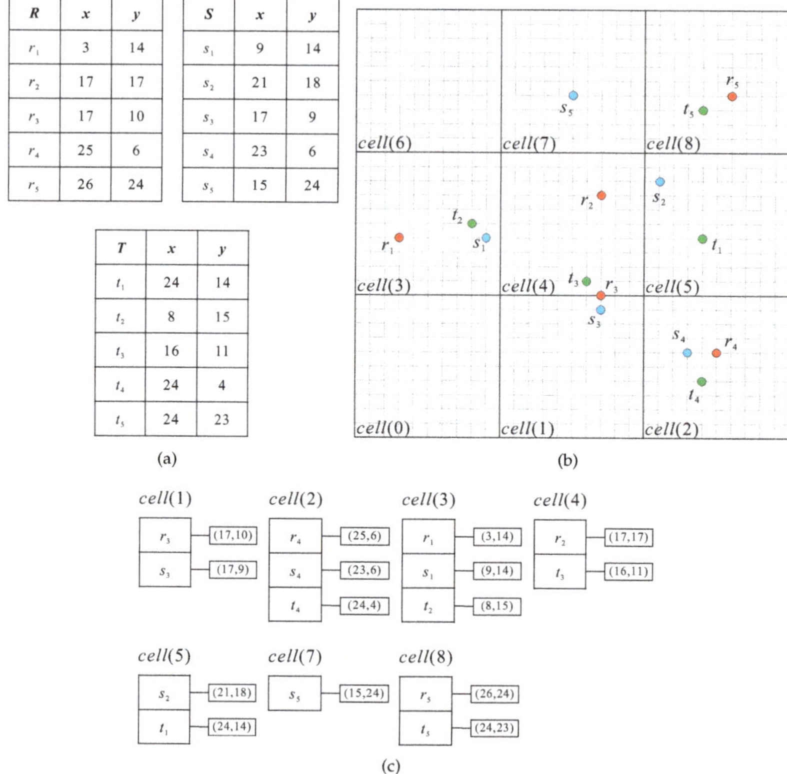

R	x	y
r_1	3	14
r_2	17	17
r_3	17	10
r_4	25	6
r_5	26	24

S	x	y
s_1	9	14
s_2	21	18
s_3	17	9
s_4	23	6
s_5	15	24

T	x	y
t_1	24	14
t_2	8	15
t_3	16	11
t_4	24	4
t_5	24	23

(a)

(b)

cell(1)
- r_3 — (17,10)
- s_3 — (17,9)

cell(2)
- r_4 — (25,6)
- s_4 — (23,6)
- t_4 — (24,4)

cell(3)
- r_1 — (3,14)
- s_1 — (9,14)
- t_2 — (8,15)

cell(4)
- r_2 — (17,17)
- t_3 — (16,11)

cell(5)
- s_2 — (21,18)
- t_1 — (24,14)

cell(7)
- s_5 — (15,24)

cell(8)
- r_5 — (26,24)
- t_5 — (24,23)

(c)

Figure 2. Illustration of grid structure and HDFS. (**a**) Heterogeneous objects; (**b**) Grid structure; (**c**) Data on HDFS.

In order to provide parallel processing of the heterogeneous objects using MapReduce, information of the grid structure is stored in a distributed storage system, the HDFS, by default. The HDFS consists of multiple DataNodes for storing data and a NameNode for monitoring all DataNodes. In the HDFS, a file is broken into multiple equal-sized chunks and then the NameNode allocates the data chunks among the DataNodes for query processing. Returning to the example in Figure 2, the cells, $cell(0)$ to $cell(8)$, are treated as the chunks and kept on the HDFS. Take the cell $cell(1)$ as an example, as objects r_3 and s_3 are enclosed in $cell(1)$, in the HDFS, the chunks with respect to $cell(1)$ will store r_3 and s_3 with their coordinates $(17, 10)$ and $(17, 9)$, respectively. Note that the cells $cell(0)$ and $cell(6)$ need not be kept on the HDFS because there is no object in them. Figure 2c shows how the grid structure for the heterogeneous objects is stored on the HDFS.

4. Mapreduce-Based Aggregate Query Algorithm

Given the n types of data objects, O_1, O_2, ..., O_n, a set of query points Q (where a query point $q \in Q$ corresponds to a *SAvgDQ*, a *SMinDQ*, a *SMaxDQ*, or a *SSumDQ*), and the user-defined distance d, the main goal of the MapReduce-based aggregate query (MRAggQ) algorithm is to efficiently determine, for each query point q, the *HNO set* with the shortest distance in a distributed manner. Recall that a set of objects $\{o_1, o_2, ..., o_n\}$ (where $o_i \in O_i$ and $i = 1 \sim n$) can be included in the result set of the location-based aggregate queries only if the following two conditions hold: (1) the distance between any two objects in $\{o_1, o_2, ..., o_n\}$ is less than or equal to d (as a necessary condition) and (2) $\{o_1, o_2, ..., o_n\}$ has the shortest average, minimal, maximal, or sum distance to the query point. As a result, the MRAggQ algorithm is developed according to the two conditions. The proposed MRAggQ algorithm consists of four phases, in which the first and last two phases are in charge of checking the conditions (1) and (2), respectively. In the following, we briefly describe the purposes of the four phases and then discuss the details separately. To provide an overview of the MRAggQ algorithm, a flowchart and a pseudo code for the four phases are also given in Figure 3 and Algorithm 1, respectively:

- The first phase, the *Inner HNO set determining phase*, aims at finding, for each cell $cell(c)$, the sets of objects that are enclosed in $cell(c)$ and are within the distance d from each other. Here, we term the object sets found in this phase the *Inner HNO sets*.
- The second phase, the *Outer HNO set determining phase*, focuses on finding the *HNO sets* that have not been discovered from the previous phase. It means that the objects constituting a *HNO set* determined in this phase cannot be fully enclosed in a cell. Instead, the objects are distributed over different cells. We term the *HNO sets* discovered in this phase the *Outer HNO sets*.
- The third phase, the *Aggregate-distance computing phase*, is responsible for computing the aggregate-distances of all *HNO sets* obtained from the previous two phases to each query point contained in the query set Q, according to the type of location-based aggregate queries (i.e., the aggregate-distance may be the average, the minimal, the maximal, or the sum distance).
- The last phase, the *Result set generating phase*, sorts the aggregate-distances of all *HNO sets* computed in the previous phase, so as to output the *HNO set* with the shortest aggregate-distance for each query point in Q.

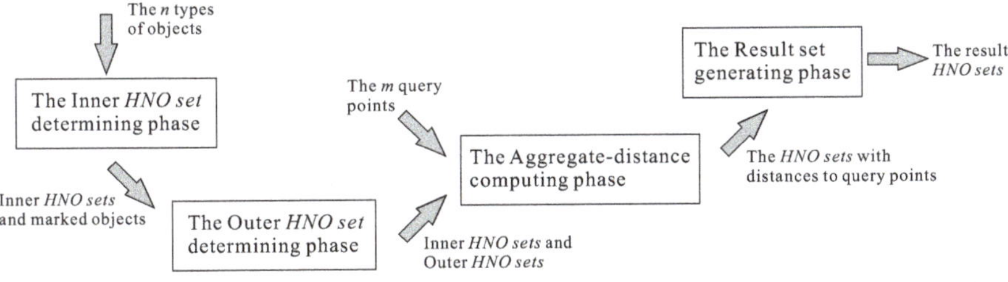

Figure 3. Flowchart of the MRAggQ algorithm.

Algorithm 1: The MRAggQ algorithm

Input : The n types of objects, and the set of m query points
Output: The result *HNO set* for each query point

```
/* The Inner HNO set determining phase                                                    */
finding the Inner HNO sets enclosed in cell(c);
determining the marked objects for cell(c);
/* The Outer HNO set determining phase                                                    */
finding the Outer HNO sets based on the marked objects;
combing the Inner and the Outer HNO sets;
/* The Aggregate-distance computing phase                                                 */
computing the average, min, max, or sum distances of the HNO sets to the m query points;
/* The Result set generating phase                                                        */
sorting the HNO sets according to their distances to each query point;
returning the HNO set with the shortest distance to each query point;
```

4.1. Inner HNO Set Determining Phase

Given the n types of objects stored on the HDFS, the goal of the Inner *HNO set* determining phase is to process in parallel, determining the Inner *HNO sets* for each cell $cell(c)$, each of which is composed of n types of objects enclosed in $cell(c)$. In this phase, a MapReduce job consisting of the *map* step, the *shuffle* step, and the *reduce* step is executed to finish the procedure. In the map step, each cell in the form of $< cell(c), \{o_i, (x_i, y_i)\} >$ (i.e., $< key, value >$ pair) is extracted from the HDFS as input. The pair $< cell(c), \{o_i, (x_i, y_i)\} >$ generated by the map step is then transmitted to another machine in the shuffle step, where the recipient machine is determined solely by value of $cell(c)$. That is, if the pairs have a common key $cell(c)$, all of them will arrive at an identical machine for processing in the reduce step. This is because for the n pairs $< cell(c), \{o_i, (x_i, y_i)\} >$ (where $i = 1 \sim n$) with the same key $cell(c)$, a set composed of the n objects $o_1, o_2, ..., o_n$ has a chance to be the Inner *HNO set* as all the objects are enclosed in the cell $cell(c)$. In the reduce step, two processing tasks are carried out in each participating machine, by taking into account the key-value pairs received from the shuffle step.

- The first task is to compute the distance between any two objects o_i and o_j enclosed in $cell(c)$, where $1 \le i, j \le n$ and $i \ne j$, based on their coordinates (x_i, y_i) and (x_j, y_j). Consider a set of objects $\{o_1, o_2, ..., o_n\}$ enclosed in the cell $cell(c)$. If the computed distances of all object pairs are less than or equal to the distance d, then $\{o_1, o_2, ..., o_n\}$ is an Inner *HNO set* of $cell(c)$. Hence, a key-value pair in the form of $< cell(c), \{\{o_1, (x_1, y_1)\}, \{o_2, (x_2, y_2)\}, ..., \{o_n, (x_n, y_n)\}\} >$ is returned as output.
- The second task, as a preliminary to the next phase, the Outer *HNO set* determining phase, focuses on marking some objects enclosed in cell $cell(c)$ that may constitute an Outer *HNO set* with the other objects enclosed in different cells. We term the objects determined by the second task the *marked objects*. For an object o_i enclosed in $cell(c)$, it can be the marked object only if the circle centered at o_i with radius d is not fully contained in $cell(c)$. Otherwise (i.e., the circle is enclosed by $cell(c)$), there exists no object enclosed in another cell $cell(c')$ and whose distance to object o_i is less than or equal to d, and thus o_i must not be contained in the Outer *HNO sets*. Suppose that the data space is divided into $C \times C$ cells, where each equal-sized cell is represented as a rectangle with widths w_x and w_y on the x-axis and y-axis, respectively. An object o_i with coordinates (x_i, y_i) is a marked object in cell $cell(c)$ if the following condition holds:

$$[x_i - d, x_i + d] \not\subseteq [(c \bmod C) \times w_x, ((c \bmod C) + 1) \times w_x], \tag{1}$$
$$[y_i - d, y_i + d] \not\subseteq [\lfloor \tfrac{c}{C} \rfloor \times w_y, (\lfloor \tfrac{c}{C} \rfloor + 1) \times w_y].$$

Similar to the first task, a key-value pair with respect to each marked object o_i (i.e., $< key_i, \{o_i, (x_i, y_i)\} >$) will be generated after executing the second task. The generated key is mainly used to guarantee that the n types of objects constituting an Outer *HNO set* can be processed in the same machine. Note that, if such objects are considered in different machines, some of the Outer *HNO sets* may be lost. In order to give each marked object o_i enclosed in the cell $cell(c)$ a key key_i, we first merge $C_x \times C_y$ cells into a rectangle R bounding the cell $cell(c)$, where the parameters C_x and C_y are estimated based on the following equation:

$$C_x = \left\lceil \frac{d}{w_x} \right\rceil + 1, C_y = \left\lceil \frac{d}{w_y} \right\rceil + 1. \tag{2}$$

Then, the key of the marked object o_i is set to the union of the ids of these cells. To establish better understanding of the main idea behind Equation (2), we take the cell $cell(4)$ in Figure 2b as an example, where the user-defined distance $d = 2.5$ and both the widths w_x and w_y of each cell are equal to 10. Based on Equation (2), a rectangle R consisting of 2×2 cells is constructed to enclose the cell $cell(4)$ (here, R can be represented as $cell(0,1,3,4)$, $cell(1,2,4,5)$, $cell(3,4,6,7)$, and $cell(4,5,7,8)$). Let us consider the rectangle R corresponding to $cell(0,1,3,4)$. As the minimal distance between $cell(4)$ and each of the other three cells, $cell(0)$, $cell(1)$, and $cell(3)$ is less than or equal to d, it is possible that an Outer *HNO set* is composed of one or more marked objects in $cell(4)$ and the rest in the other three cells. As such, we should give all the marked objects enclosed in the rectangle R a common key, $cell(0,1,3,4)$, so as to process them in the same machine. In addition, the keys $cell(1,2,4,5)$, $cell(3,4,6,7)$, and $cell(4,5,7,8)$ are assigned to the marked objects enclosed in their corresponding rectangle R in the same way.

Figure 4 is a concrete example, which continues the previous example in Figure 2, illustrating the data flow of the MapReduce job for the Inner *HNO set* determining phase. In the map step, a key $cell(c)$ for each object o_i is extracted and transformed into key-value pair, $< cell(c), \{o_i, (x_i, y_i)\} >$ (e.g., $< cell(3), \{r_1, (3, 14)\} >$ for the object r_1). Then, the key-value pairs with the same key are shuffled to the same machine for processing. For example, $< cell(1), \{r_3, (17, 10)\} >$ and $< cell(1), \{s_3, (17, 9)\} >$ arrive at the same machine because of their common key $cell(1)$. In the reduce step, each participating machine carries out the first task (i.e., determining the Inner *HNO sets*) by computing the distance between objects received from the shuffle step to compare with the distance $d = 2.5$. In this figure, $< cell(2), \{\{r_4, (25, 6)\}, \{s_4, (23, 6)\}, \{t_4, (24, 4)\}\} >$ is output, so that $\{r_4, s_4, t_4\}$ is an Inner *HNO set* enclosed in the cell $cell(2)$. Meanwhile, the second task (i.e., finding the marked objects) is executed in each machine to find the marked objects enclosed in a cell based on Equation (1) and give each marked object a key according to Equation (2). Take the marked object t_3 enclosed in the cell $cell(4)$ as an example. Four key-value pairs, $< cell(0,1,3,4), \{t_3, (16, 11)\} >$, $< cell(1,2,4,5), \{t_3, (16, 11)\} >$, $< cell(3,4,6,7), \{t_3, (16, 11)\} >$, and $< cell(4,5,7,8), \{t_3, (16, 11)\} >$ will be output from the machine in charge of $cell(4)$, as there is a chance that t_3 constitutes an Outer *HNO set* with the other marked objects enclosed in its surrounding cells. Finally, the Inner *HNO sets* and the marked objects with respect to each cell, discovered in the Inner *HNO set* determining phase, are passed to the next phase.

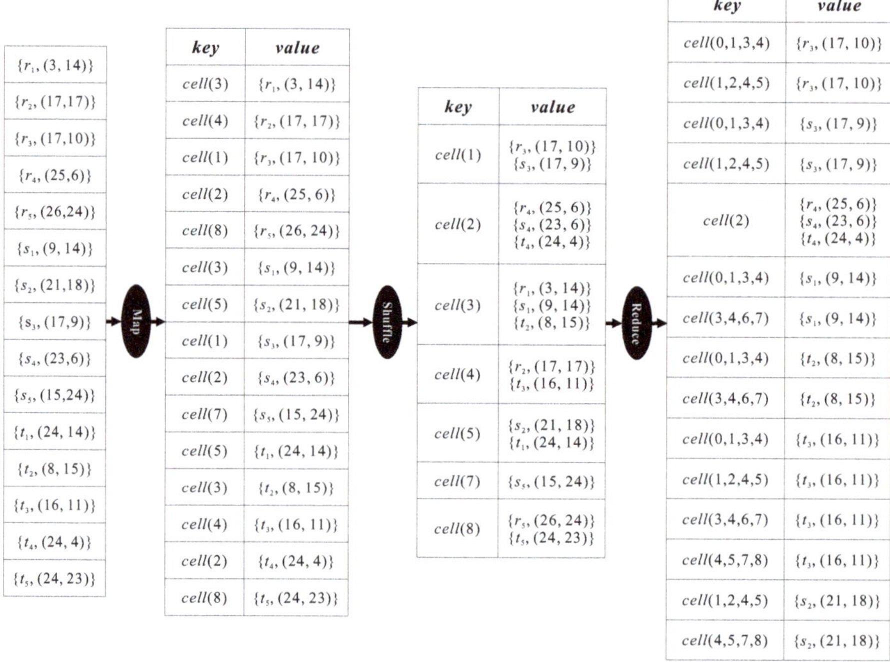

Figure 4. Illustration of the Inner *HNO set* determining phase.

4.2. Outer HNO Set Determining Phase

The Outer *HNO set* determining phase focuses on finding the *HNO sets* that have not been discovered (i.e., the Outer *HNO sets*), by exploiting information of the marked objects obtained from the previous phase. Similarly, a MapReduce job is applied in the Outer *HNO set* determining phase, where (1) the map step receives the result of the previous phase and the key-value pairs are emitted, (2) the shuffle step dispatches the pairs with the same key to an identical machine for checking whether the Outer *HNO sets* exist, and (3) the reduce step computes the distance between the marked objects to compare with the distance d. Having executed the Outer *HNO set* determining phase, each key-value pair in the form of $< cell(c), \{\{o_1, (x_1, y_1)\}, \{o_2, (x_2, y_2)\}, ..., \{o_n, (x_n, y_n)\}\} >$ is returned as output, where c refers to either a cell id (meaning that $\{\{o_1, (x_1, y_1)\}, \{o_2, (x_2, y_2)\}, ..., \{o_n, (x_n, y_n)\}\} >$ is an Inner *HNO set*) or multiple cell ids (that is, an Outer *HNO set*). Continuing the example in Figure 4, the key-value pairs corresponding to an Inner *HNO set*, $< cell(2), \{\{r_4, (25,6)\}, \{s_4, (23,6)\}, \{t_4, (24,4)\}\} >$, and the marked objects, $< cell(0,1,3,4), \{r_3, (17,10)\} >$ and so on, are emitted in the map step of the Outer *HNO set* determining phase, as shown in Figure 5. In the shuffle step, the marked objects with the common key are assigned to the same machine for computing the distance between any two marked objects based on their coordinates. For instance, five marked objects r_3, s_3, t_1, t_2, and t_3 with the key $cell(0,1,3,4)$ will be considered in the same machine. In the reduce step, each participating machine computes the distance between the marked objects assigned by the shuffle step (note that only the distances between different types of marked objects are computed), and then outputs the Outer *HNO sets*. In this figure, the key-value pairs, $< cell(0,1,3,4), \{\{r_3, (17,10)\}, \{s_3, (17,9)\}, \{t_3, (16,11)\}\} >$ and $< cell(1,2,4,5), \{\{r_3, (17,10)\}, \{s_3, (17,9)\}, \{t_3, (16,11)\}\} >$ are returned as they satisfy the constraint of distance d. As we can see, $\{\{r_3, (17,10)\}, \{s_3, (17,9)\}, \{t_3, (16,11)\}$ is a duplicate set and needs to be eliminated. The duplicate elimination will be carried out in the last phase, the Result set generating phase.

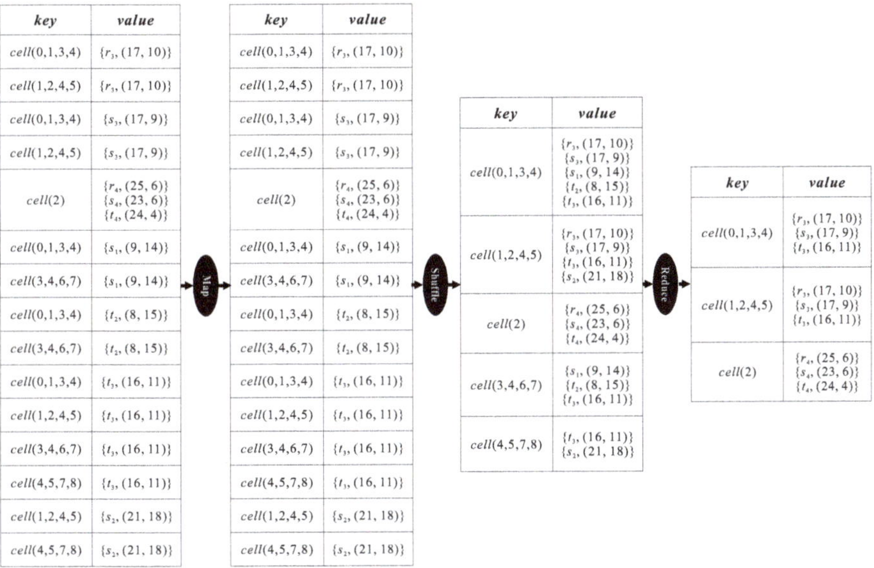

Figure 5. Illustration of the Outer *HNO set* determining phase.

4.3. Aggregate-Distance Computing Phase

After executing the first two phases (i.e., the Inner *HNO set* determining phase and the Outer *HNO set* determining phase), all of the *HNO sets* in the space can be discovered in a distributed manner. In the sequel, the third phase, the Aggregate-distance computing phase, is designed to compute in parallel the aggregate-distance of each *HNO set* according to the type of location-based aggregate queries. Suppose that Q is a set of m query points, q_1, q_2, ..., q_m, at which a *SAvgDQ*, a *SMinDQ*, a *SMaxDQ*, or a *SSumDQ* is issued. A query table with respect to Q needs to be broadcast to each machine so as to estimate the aggregate-distances between the *HNO sets* processed by this machine and each query point in Q. Each tuple of the query table has two fields: the query id q_i^j (where j can be 1, 2, 3, and 4, indicating *SAvgDQ*, *SMinDQ*, *SMaxDQ*, and *SSumDQ*, respectively) and the coordinates (x_{q_i}, y_{q_i}). In the map step of the Aggregate-distance computing phase, in addition to the key-value pair $< cell(c), \{\{o_1, (x_1, y_1)\}, \{o_2, (x_2, y_2)\}, ..., \{o_n, (x_n, y_n)\}\} >$ for each *HNO set*, a key-value pair $< cell(c), \{\{q_1^j, (x_{q_1}, y_{q_1}), v_1\}, \{q_2^j, (x_{q_2}, y_{q_2}), v_2\}, ..., \{q_m^j, (x_{q_m}, y_{q_m}), v_m\}\} >$ with regard to the query points is also emitted, so that the query set Q can be transmitted along with each *HNO set* to the same machine for query processing. Having executed the shuffle step, the *HNO set* $\{o_1, o_2, ..., o_n\}$ and the query set $\{q_1, q_2, ..., q_m\}$ with the same key $cell(c)$ are grouped together. For each participating machine, the task of computing the aggregate-distance between each *HNO set* and each query point assigned by the shuffle step is carried out in the reduce step, in which the aggregate-distance refers to the average, minimal, maximal, or sum distance according to the query type (i.e., the value of j). Finally, each key-value pair in the form of $< q_i^j, \{(o_1, o_2, ..., o_n), d_{agg}\} >$ is returned as output, where d_{agg} is the aggregate-distance between the *HNO set* $\{o_1, o_2, ..., o_n\}$ and the query point q_i.

As shown in Figure 6, continuing the example of Figure 5, the query table maintains four query points q_1 to q_4 with their coordinates and query types, in which q_1^1, q_2^2, q_3^3, and q_4^4 issue the *SAvgDQ*, the *SMinDQ*, the *SMaxDQ*, and the *SSumDQ*, respectively. In the map step, the key-value pairs $< cell(0, 1, 3, 4), \{\{r_3, (17, 10)\}, \{s_3, (17, 9)\}, \{t_3, (16, 11)\}\} >$, $< cell(1, 2, 4, 5), \{\{r_3, (17, 10)\}, \{s_3, (17, 9)\}, \{t_3, (16, 11)\}\} >$, and $< cell(2), \{\{r_4, (25, 6)\}, \{s_4, (23, 6)\}, \{t_4, (24, 4)\}\} >$ obtained from the previous phase (i.e., the Outer *HNO set* determining phase) are emitted. For the sake of grouping the *HNO sets* and the query points, the key-value pairs, $< cell(0, 1, 3, 4), \{\{q_1^1, (26, 4)\}, \{q_2^2, (6, 17)\},$

$\{q_3^3, (17, 21)\}, \{q_4^4, (21, 9)\}\} >$ and so on, are also generated based on the keys $cell(0, 1, 3, 4)$, $cell(1, 2, 4, 5)$, and $cell(2)$. The shuffle step dispatches the pairs with the same key to the same machine for computing the aggregate-distance between the *HNO sets* and the query points. Take the key-value pair with regard to the key $cell(0, 1, 3, 4)$ as an example. The machine in charge of $cell(0, 1, 3, 4)$ runs the reduce step to compute the average distance of the *HNO set* $\{r_3, s_3, t_3\}$ to the query point q_1 as the query type $j = 1$. Then, a key-value pair $< q_1^1, \{(r_3, s_3, t_3), 11.1\} >$ is output, meaning that the average distance is equal to 11.1. Similarly, in the reduce step, the min, max, and sum distances of $\{r_3, s_3, t_3\}$ to q_2, q_3, and q_4, are estimated as 11.66, 12, and 6.41, respectively. After the key-value pairs have been output by all the participating machines, the last phase, the Result set generating phase, will sort the *HNO sets* in ascending order of their aggregate-distance to determine the query result.

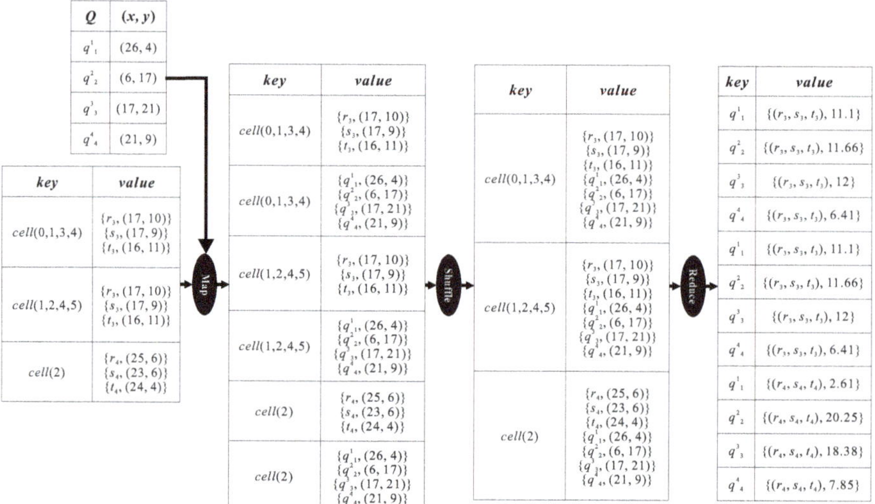

Figure 6. Illustration of the Aggregate-distance computing phase.

4.4. Result Set Generating Phase

The goal of the last phase, the Result set generating phase, is to determine the *HNO set* with the shortest aggregate-distance for each query point in a distributed manner. Once a MapReduce job starts, the key-value pairs $< q_i^j, \{(o_1, o_2, ..., o_n), d_{agg}\} >$ received from the previous phase are directly emitted in the map step. According to the key q_i^j, the *HNO sets* having the same q_i^j will be assigned to an identical machine in the shuffle step because their aggregate-distances to the query point q_i need to be compared so as to determine the query result for q_i. For the machine receiving the key-value pairs with respect to q_i, the first task of the reduce step is to eliminate the duplicate value in the form of $\{(o_1, o_2, ..., o_n), d_{agg}\}$. Then, the second task is to sort the *HNO sets* in ascending order of their aggregate-distance d_{agg}, and finally output the *HNO set* with smallest d_{agg} as the query result.

Figure 7 gives an illustration of how the Result set generating phase is executed using the key-value pairs generated from the previous phase (shown in Figure 6). In the map step, all key-value pairs which use the query id as the key (e.g., $< q_1^1, \{(r_3, s_3, t_3), 11.1\} >$) are emitted so that the pairs with the same key can be grouped together for processing after the shuffle step. For instance, the pairs $< q_1^1, \{(r_3, s_3, t_3), 11.1\} >$ and $< q_1^1, \{(r_4, s_4, t_4), 2.61\} >$ are assigned to the same machine because of their common key q_1^1. By executing the reduce step in each machine, the duplicates are first removed and then the *HNO set* with the shortest aggregate-distance for each query point is output. In this figure, the *HNO set* $\{r_4, s_4, t_4\}$ is the result for the query point q_1, and the *HNO set* $\{r_3, s_3, t_3\}$ is the result for the other three query points q_2, q_3, and q_4.

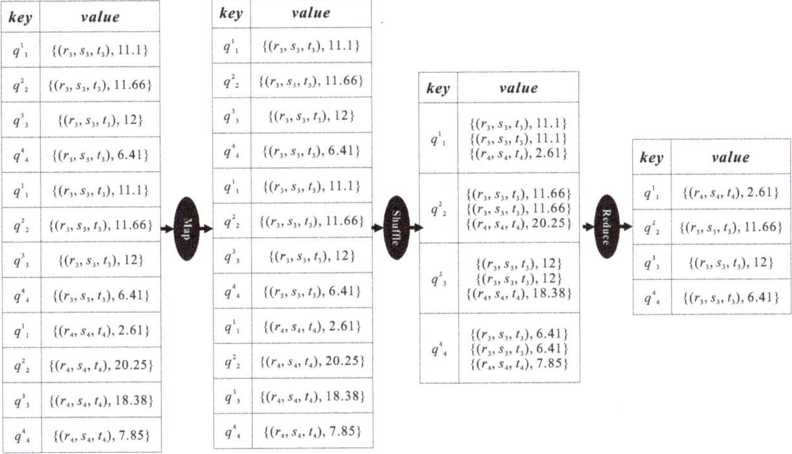

Figure 7. Illustration of the Result set generating phase.

5. Performance Evaluation

In this section, we conduct a series of experiments to evaluate the performance of the proposed *MRAggQ* algorithm. We first study the effect of the used grid structure on the performance of processing the location-based aggregate queries, so as to decide an appropriate number of heterogeneous objects enclosed in each cell for query processing. Then, we demonstrate the efficiency and the scalability of the proposed *MRAggQ* algorithm by measuring its running time with respect to various important factors.

5.1. Performance Settings

All of the experiments are performed on a cluster of four computing machines. Each computing machine is a PC with Intel 2.70 GHz CPU and 16 GB RAM, and runs 32-bit Ubuntu 15.10. The algorithms are implemented in JAVA and allocate 4 GB RAM to the Java Virtual Machine. The computing machines are connected by a 1000 Mbps Ethernet and Hadoop 2.6.0 is used as the default distributed file system. One synthetic dataset and four real datasets are evaluated in our simulation. The synthetic dataset consists of n types of objects, where n varies from 1 to 5, and the total number of objects ranges from 1000 K to 5000 K. The objects are spread over a region of $1,000,000 \times 1,000,000$ with the *Uniform distribution*. As for the real datasets, we consider the *Beijing, Manchester, Pittsburgh,* and *Charlotte* files (containing about 400 K, 1000 K, 1200 K, and 1500 K objects, respectively) extracted from the *OpenStreetMap* [38]. The data space is divided into multiple cells by considering the parameter α (i.e., the maximum number of objects enclosed in a cell), which varies from 250 to 2000 in the experiments, and then a grid structure is built to manage the heterogeneous objects enclosed in the cells. In the experimental space, we also generate a set of m query points (where m varies from 0.1 K to 10 K). Each of the query points issues a *SAvgDQ*, a *SMinDQ*, a *SMaxDQ*, or a *SSumDQ* to the server, where the user-defined distance d ranges from 0.1% to 2% of the width of the cell.

The performance of the *MRAggQ* algorithm is measured by the average running time in performing workloads of the location-based aggregate queries issued from the m query points. We investigate the effects of five important factors on the performance of processing the location-based aggregate queries, including *the parameter α* (used for the grid index), *the number of objects, the number of object types* (i.e., n), *the user-defined distance d,* and *the number of query points.* Table 1 summarizes the parameters under investigation, along with their default values and ranges. In the sequel, we show the experimental results with detailed discussions for these important factors, respectively.

Table 1. System parameters.

Parameter	Default	Range
Parameter α	1000	250, 500, 1000, 1500, 2000
Number of objects (K)	1000	1000, 2000, 3000, 4000, 5000
Number of object types	3	1, 2, 3, 4, 5
Distance d (%)	1	0.1, 0.5, 1, 1.5, 2
Number of query points (K)	1	0.1, 0.5, 1, 5, 10

5.2. Effect of Parameter α

The first set of experiments studies the effect of the number of objects enclosed in each cell (i.e., the parameter α) on the performance of processing the location-based aggregate queries, using the Uniform dataset and the Manchester dataset. In the experiments, we vary the value of the parameter α from 250 to 2000 and evaluate the average running time for the proposed *MRAggQ* algorithm. For both the Uniform dataset and the Manchester dataset, the average running time first decreases and then increases with the increasing value of α, as shown in Figure 8a,b, respectively. This is mainly because (1) for a smaller α (i.e., fewer number of objects in each cell), more cells need to be generated for storing object information, and thus each participating machine (i.e., the DataNode) spends more time on processing the increasing number of cells assigned by the NameNode, while (2) for a greater α (meaning that the number of cells decreases but the storage overhead for each cell increases), computing the distances between objects to determine the Inner and the Outer *HNO sets* with respect to each cell takes more processing time. As the parameter α dominates the performance of processing the location-based aggregate queries, we need to decide an appropriate value of α used to partition the data space. As we can see, for both the Uniform dataset and the Manchester dataset, the average running time increases noticeably after $\alpha = 1000$. The experimental result shows that $\alpha = 1000$ is a better choice than the others, and hence will be used as the default value in all the rest experiments.

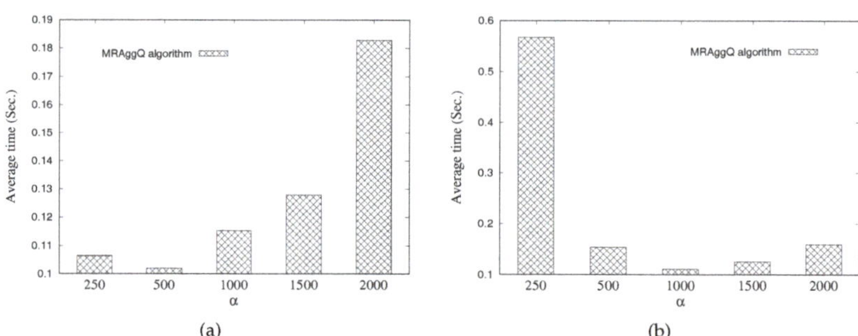

Figure 8. Effect of the parameter α. (**a**) Uniform dataset; (**b**) Manchester dataset.

5.3. Effect of Number Of Objects

The second set of experiments illustrates the performance of processing the location-based aggregate queries using the Uniform dataset (in which the number of objects varies from 1000 K to 5000 K) and the real dataset (including the *Beijing, Manchester, Pittsburgh,* and *Charlotte* files). As shown in Figure 9a, the average running time for the *MRAggQ* algorithm increases with the increasing number of objects. The reason is that a larger number of objects results in more cells to be processed, so that a majority of the running time is spent on executing the Inner *HNO set* determining phase and the Outer *HNO set* determining phase. Nevertheless, benefited from processing the location-based aggregate queries in a distributed manner, the average running time for all cases remains below 0.25 s. As for the real dataset, shown in Figure 9b, the *Beijing* file contains fewer objects than the *Manchester*, *Pittsburgh*, and *Charlotte* files, but incurs the highest average running time. This is because the *Beijing* file has a denser object distribution (compared to the other three files), thus leading to more *HNO sets* to be considered in the Aggregate-distance computing phase and the Result set generating phase.

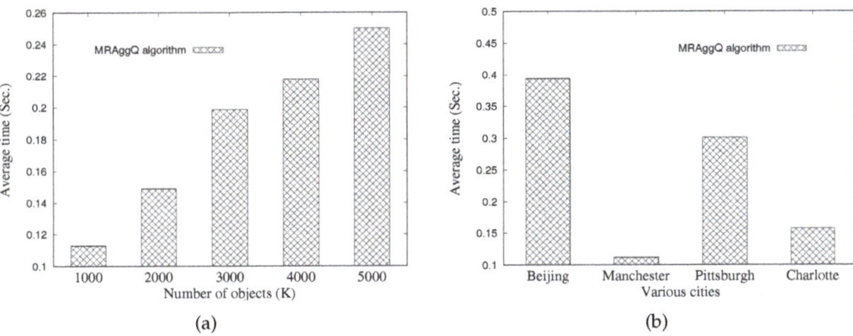

Figure 9. Effect of the number of objects. (**a**) Uniform dataset; (**b**) Real dataset.

5.4. Effect of Number of Object Types

The third set of experiments is conducted to investigate the impact of the number of object types (i.e., the value of n) on the performance of the *MRAggQ* algorithm. Figure 10a,b measure the average running time of the *MRAggQ* algorithm for the Uniform dataset and the Manchester dataset, respectively, by varying n from 1 to 5. In the case where $n = 1$, implying that only single type of objects is considered, the processing cost required for determining whether the road distance between different types of objects exceeds the distance d can be completely avoided (that is, the first two phases of the *MRAggQ* algorithm do nothing). Moreover, the problem of processing the *SAvgDQ*, the *SMinDQ*, the *SMaxDQ*, and the *SSumDQ* is reduced to finding the nearest neighbor of the query object (i.e., the object with the shortest distance). In the case that n gets larger than 1, the Inner *HNO set* determining phase and the Outer *HNO set* determining phase need to be executed to find the *HNO sets*, as more than one type of object is processed. This is why the average running time of processing the location-based aggregate queries grows as the value of n increases. The experimental results also show that (1) the nearest neighbor query is a special case of location-based aggregate queries, where $d = \infty$ and $n = 1$, and (2) the proposed *MRAggQ* algorithm can be successfully applied to process the nearest neighbor query in a distributed manner.

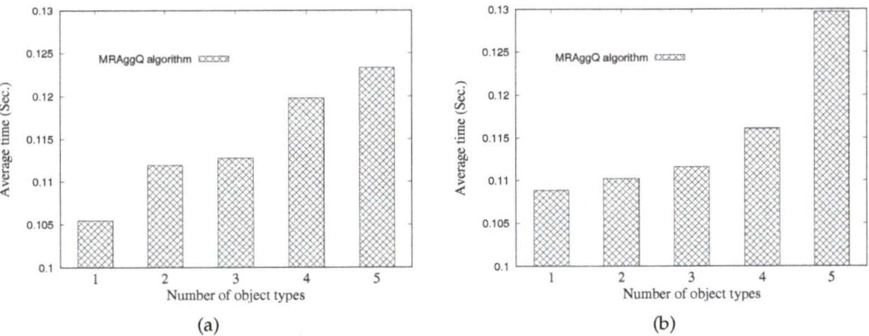

Figure 10. Effect of the number of object types. (**a**) Uniform dataset; (**b**) Manchester dataset.

5.5. Effect of Distance D

The fourth set of experiments illustrates the average running time of the *MRAggQ* algorithm as a function of the distance d (ranging from 0.1% to 2.0% of the width of the cell). As shown in Figure 11a,b, for both the Uniform dataset and the Manchester dataset, the larger the distance d, the higher the overhead in processing the location-based aggregate queries. This is attributed to the fact that, (1) for a smaller distance d, most of the object pairs cannot satisfy the constraint of distance d, reducing a lot of redundant distance computations in determining whether an object set can be the *HNO set* or not, and (2) a larger distance d makes the distance constraint easier to be satisfied for each object pair, incurring higher cost of computing object distances (in the Inner *HNO set* determining phase and the Outer *HNO set* determining phase) and more *HNO sets* to be considered (in the Aggregate-distance computing phase and the Result set generating phase).

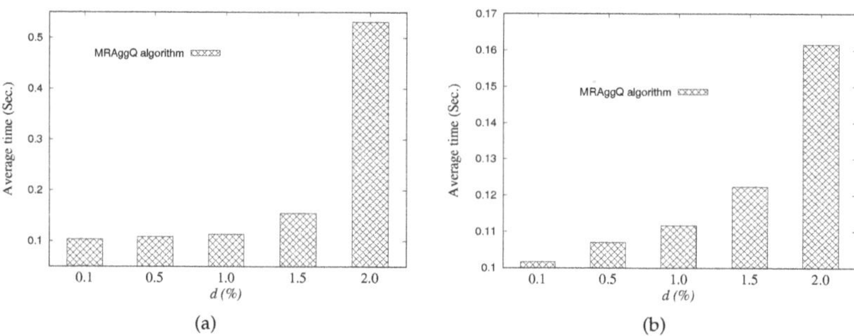

Figure 11. Effect of the distance d. (**a**) Uniform dataset; (**b**) Manchester dataset.

5.6. Effect of Number of Query Points

As the main focus of this paper is to process multiple location-based aggregate queries in a distribute manner, the *scalability* is the most important performance metric for the proposed *MRAggQ* algorithm—that is, the average CPU time it takes for the *MRAggQ* algorithm to simultaneously process multiple location-based aggregate queries. Therefore, in this subsection, a centralized approach proposed in the previous work [7], termed the *centralized algorithm*, is used as a competitor. The set of experiments demonstrates the scalability of the *MRAggQ* algorithm, compared to the centralized algorithm, by studying the effect of the number of location-based aggregate queries in terms of the average running time. Figure 12a,b evaluate the processing overhead for the Uniform dataset and the Manchester dataset, respectively, under various numbers of query points (varying from 0.1 K to 10 K). Note that the two figures use a logarithmic scale for the y-axis. As we can see, both curves for the centralized algorithm exhibit the increasing trends with the increase of number of queries. This is because, in the centralized algorithm, the location-based aggregate queries need to be processed sequentially. It is very likely that the overall performance suffers from some time-consuming queries, and hence the average CPU time drastically increases for a large number of queries. Conversely, an interesting observation from the experimental results is that a larger number of queries results in a lower average running time for the *MRAggQ* algorithm. The reason for this improvement is that the increasing number of queries imposes only the computational burden on the last two phases (i.e., the Aggregate-distance computing phase and the Result set generating phase), instead of all the four phases of the *MRAggQ* algorithm. As a result, processing more queries would reduce more running time from the average perspective. The experimental results also show that there is a wide gap between the *MRAggQ* algorithm and the centralized algorithm, which confirms that the *MRAggQ* algorithm is scalable and efficient in highly dynamic environments where multiple location-based aggregate queries have to be processed concurrently.

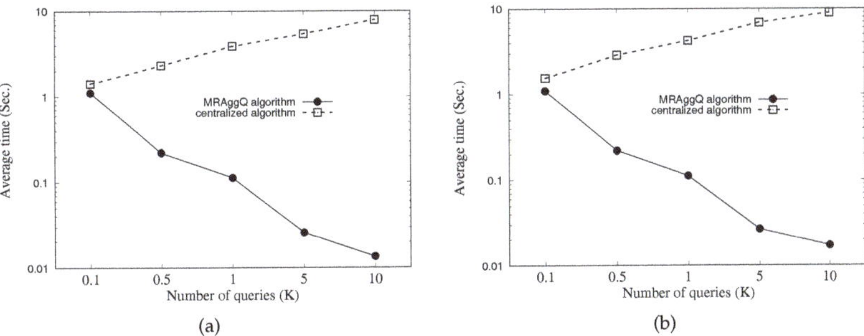

Figure 12. Effect of the number of queries. (**a**) Uniform dataset; (**b**) Manchester dataset.

6. Conclusions

In this paper, we focus on efficiently processing multiple location-based aggregate queries (including the *SAvgDQ*, the *SMinDQ*, the *SMaxDQ*, and the *SSumDQ*) in a distributed manner. We adopt the most notable MapReduce platform for parallelizing location-based aggregate query processing. A grid structure is first utilized to manage the heterogeneous objects by taking into account the storage balance, and then the *MRAggQ* algorithm is developed based on MapReduce, in which four MapReduce jobs are executed, to provide distributed processing of multiple location-based aggregate queries.

There are several interesting avenues for the future extensions of this work. One important research direction is to further improve the overall performance of the *MRAggQ* algorithm. For a heavy query load, the different keys may not be processed by different computing machines in parallel, thus decreasing the query performance. Therefore, our next step is to enhance the query performance by considering different clusters and/or Hadoop configurations (e.g., applying the SpatialHadoop). Then, an important avenue is to extend the distributed processing technique to be suitable for the environment where the heterogeneous objects move with time. Moreover, an extension is to address the issue of modifying the distributed processing technique to answer the location-based aggregate queries in road networks. Finally, a further extension is how to answer other variations of the location-based aggregate queries using a MapReduce platform.

Funding: This work was supported by the Ministry of Science and Technology of Taiwan under Grant MOST 107-2119-M-992-304 and Grant MOST 108-2621-M-992-002.

Conflicts of Interest: The author declares no conflict of interest.

References

1. Benetis, R.; Jensen, C.S.; Karciauskas, G.; Saltenis, S. Nearest Neighbor and Reverse Nearest Neighbor Queries for Moving Objects. *VLDB J.* **2006**, *15*, 229–249. [CrossRef]
2. Chen, J.; Cheng, R. *Efficient Evaluation of Imprecise Location-Dependent Queries*; ICDE: Oslo, Norway, 2007; pp. 586–595.
3. Mokbel, M.F.; Xiong, X.; Aref, W.G. SINA: Scalable Incremental Processing of Continuous Queries in Spatio-temporal Databases. In Proceedings of the 2004 ACM SIGMOD, Paris, France, 13–18 June 2004; pp. 623–634.
4. Sharifzadeh, M.; Shahabi, C. The Spatial Skyline Queries. In Proceedings of the International Conference on Very Large Data Bases, Seoul, Korea, 12–15 September 2006; pp. 751–762.
5. Sistla, A.P.; Wolfson, O.; Chamberlain, S.; Dao, S. *Modeling and Querying Moving Objects*; ICDE: Oslo, Norway, 1997; pp. 422–432.
6. Tao, Y.; Papadias, D. Time-parameterized queries in spatio-temporal databases. In Proceedings of the 2002 ACM SIGMOD, Madison, WI, USA, 3–6 June 2002; pp. 334–345.

7. Huang, Y.K. Location-Based Aggregate Queries for Heterogeneous Neighboring Objects. *IEEE Access* **2017**, *5*, 4887–4899. [CrossRef]
8. Dean, J.; Ghemawat, S. MapReduce: Simplified data processing on large clusters. In Proceedings of the 6th Symposium on Operating System Design and Implementation (OSDI 2004), San Francisco, CA, USA, 6–8 December 2004; pp. 137–150.
9. Guttman, A. R-Trees: A Dynamic Index Structure for Spatial Searching. In Proceedings of the 1984 ACM SIGMOD, Boston, MA, USA, 18–21 June 1984; pp. 47–57.
10. Papadopoulos, A.; Manolopoulos, Y. Multiple Range Query Optimization in Spatial Databases. In Proceedings of the ADBIS '98, Poznan, Poland, 7–10 September 1998; pp. 71–82.
11. Huang, Y.K.; Lin, L.F. Continuous Within Query in Road Networks. In Proceedings of the IWCMC 2011, Istanbul, Turkey, 5–8 July 2011; pp. 1176–1181.
12. Iwerks, G.; Samet, H.; Smith, K. Continuous K-Nearest Neighbor Queries for Continuously Moving Points with Updates. In Proceedings of the International Conference on Very Large Data Bases, Berlin, Germany, 9–12 September 2003; pp. 512–523.
13. Kalashnikov, D.V.; Prabhakar, S.; Hambrusch, S.; Aref, W. Efficient Evaluation of Continuous Range Queries on Moving Objects. International Conference on Database and Expert Systems Applications, Linz, Austria, 26–29 August 2002; pp. 731–740.
14. Chung, B.S.; Lee, W.C.; Chen, A.L. Processing Probabilistic Spatio-Temporal Range Queries over Moving Objects with Uncertainty. In Proceedings of the EDBT 2009, Saint-Petersburg, Russia, 23–26 March 2009; pp. 60–71.
15. Hjaltason, G.R.; Samet, H. Distance browsing in spatial databases. *ACM Trans. Database Syst.* **1999**, *24*, 265–318. [CrossRef]
16. Roussopoulos, N.; Kelley, S.; Vincent, F. Nearest neighbor queries. In Proceedings of the 1995 ACM SIGMOD, San Jose, CA, USA, 22–25 May 1995; pp. 71–79.
17. Chen, Y.; Patel, J.M. *Efficient Evaluation of All-Nearest-Neighbor Queries*; ICDE: Oslo, Norway, 2007; pp. 1056–1065.
18. Xia, C.; Lu, H.; Ooi, B.C.; Hu, J. GORDER: An Efficient Method for KNN Join Processing. In Proceedings of the VLDB 2004, Toronto, ON, Canada, 31 August–3 September 2004.
19. Papadias, D.; Shen, Q.; Tao, Y.; Mouratidis, K. *Group Nearest Neighbor Queries*; ICDE: Oslo, Norway, 2004; pp. 301–312.
20. Bentley, J.L.; Kung, H.T.; Schkolnick, M.; Thompson, C.D. On the average number of maxima in a set of vectors and applications. *J. ACM* **1978**, *25*, 536–543. [CrossRef]
21. Kung, H.T.; Luccio, F.; Preparata, F.P. On finding the maxima of a set of vectors. *J. ACM* **1975**, *22*, 469–476. [CrossRef]
22. Borzsonyi, S.; Kossmann, D.; Stocker, K. The skyline operator. In Proceedings of the 17th International Conference on Data Engineering, Heidelberg, Germany, 2–6 April 2001; pp. 421–430.
23. Hjaltason, G.R.; Samet, H. Incremental Distance Join Algorithms for Spatial Databases. In Proceedings of the International Conference on ACM SIGMOD, Seattle, WA, USA, 2–4 June 1998; pp. 237–248.
24. Brinkhoff, T.; Kriegel, H.P.; Seeger, B. Efficient Processing of Spatial Joins Using R-trees. In Proceedings of the International Conference on ACM SIGMOD, Washington, DC, USA, 26–28 May 1993; pp. 237–246.
25. Mamoulis, N.; Papadias, D. Multiway Spatial Joins. *ACM Trans. Database Syst.* **2001**, *26*, 424–475. [CrossRef]
26. Zhang, D.; Chan, C.Y.; Tan, K.L. Nearest Group Queries. In Proceedings of the International Conference on SSDBM, Baltimore, MD, USA, 29–31 July 2013.
27. Cary, A.; Sun, Z.; Hristidis, V.; Rishe, N. Experiences on processing spatial data with mapreduce. In Proceedings of the International Conference on Scientific and Statistical Database Management, New Orleans, LA, USA, 2–4 June 2009; pp. 302–319.
28. Zhang, S.; Han, J.; Liu, Z.; Wang, K.; Feng, S. Spatial queries evaluation with mapreduce. In Proceedings of the International Conference on Grid and Cooperative Computing, Lanzhou, China, 27–29 August 2009; pp. 287–292.
29. Ji, C.; Dong, T.; Li, Y.; Shen, Y.; Li, K.; Qiu, W.; Qiu, W.; Guo, M. Inverted grid-based knn query processing with mapreduce. In Proceedings of the ChinaGrid Annual Conference, Beijing, China, 20–23 September 2012; pp. 25–32.
30. Ji, C.; Hu, H.; Xu, Y.; Li, Y.; Qu, W. Efficient multi-dimensional spatial RKNN query processing with mapreduce. In Proceedings of the ChinaGrid Annual Conference, Jilin, China, 22–23 August 2013; pp. 63–68.

31. Akdogan, A.; Demiryurek, U.; Banaei-Kashani, F.; Shahabi, C. Voronoi-based geospatial query processing with mapreduce. In Proceedings of the International Conference on Cloud Computing Technology and Science, Indianapolis, IN, USA, 30 November–3 December 2010; pp. 9–16.
32. Yokoyama, T.; Ishikawa, Y.; Suzuki, Y. Processing all k-nearest neighbor queries in hadoop. In Proceedings of the International Conference on Web-Age Information Management, Harbin, China, 18–20 August 2012; pp. 346–351.
33. Zhang, C.; Li, F.; Jestes, J. Efficient Parallel kNN Joins for Large Data in MapReduce. In Proceedings of the EDBT 2012, Berlin, Germany, 27–30 March 2012.
34. Lu, W.; Shen, Y.; Chen, S.; Ooi, B.C. Efficient processing of k nearest neighbor joins using mapreduce. In Proceedings of the International Conference on Very Large Data Bases, Istanbul, Turkey, 27–31 August 2012; pp. 1016–1027.
35. Eldawy, A.; Alarabi, L.; Mokbel, M.F. Spatial Partitioning Techniques in SpatialHadoop. In Proceedings of the 41st International Conference onVery Large Data Bases, Kohala Coast, HI, USA, 31 August–4 September 2015.
36. Eldawy, A.; Mokbel, M.F. SpatialHadoop: A MapReduce Framework for Spatial Data. In Proceedings of the 2015 IEEE 31st International Conference on Data Engineering, Seoul, Korea, 13–17 April 2015 .
37. Belussi, A.; Migliorini, S.; Eldawy, A. Detecting Skewness of Big Spatial Data in SpatialHadoop. In Proceedings of the 26th ACM SIGSPATIAL, Seattle, WA, USA, 6–9 November 2018 .
38. OpenStreetMap. Available online: http://www.openstreetmap.org/ (accessed on 23 August 2019)

 International Journal of
Geo-Information

Article

Towards the Development of Agenda 2063 Geo-Portal to Support Sustainable Development in Africa

Paidamwoyo Mhangara *, Asanda Lamba, Willard Mapurisa and Naledzani Mudau

South African National Space Agency, Earth Observation Directorate, Pretoria 0087, South Africa
* Correspondence: pmangara@sansa.org.za

Received: 30 June 2019; Accepted: 30 August 2019; Published: 6 September 2019

Abstract: The successful implementation of the African Union's Agenda 2063 strategic development blueprint is critical for the attainment of economic development, social prosperity, political stability, protection, and regional integration in Africa. Agenda 2063 is a strategic and endogenous development plan that seeks to strategically and competitively reposition the African continent to ensure poverty eradication and equitable people-centric socio-economic and technological transformation. Its impact areas include wealth creation, shared prosperity, sustainable environment, and transformative capacities. Monitoring and evaluation systems play a critical role in collecting, recording, storing, integrating, and evaluating and tracking performance information in the implementation of longer-term strategic plans. The usage of the geographic information system (GIS) as a monitoring and evaluation tool has gained traction in the last few decades due to its ability to support the collection, integration, storage, analysis, output, and distribution of location-based data. The advent of web-based GIS provides a powerful online platform to collect, integrate, discover, use and share geospatial data, information, and services related to sustainable development. In this paper, we aim to describe the implementation, architectural structural design, and the functionality of the pilot Agenda 2063 geoportal. The live prototype internet-based geoportal is intended to facilitate data collection, management, integration, analysis, and visualization of Agenda 2063 development indicators. This geoportal is meant to support the planning, implementation, and monitoring of the Agenda 2063 goals at the continental, regional, and national levels. As our results show, we successfully demonstrated that a web-geoportal is a powerful interactive platform to upload, access, explore, visualize, analyse, and disseminate geospatial data related to the sustainable development of the African continent. Although in the pilot phase, the geoportal demonstrates the primary functionality of geoportals in terms of its capability to discover, analyse, share, and download geospatial datasets.

Keywords: sustainable development; Agenda 2063; geoportal; monitoring and evaluation; GIS; geospatial data

---

1. Introduction

Africa's developmental trajectory, especially its social prosperity, economic development, public security and political stability, protection, and regional integration are largely hinged on the full implementation of the Agenda 2063 strategic blueprint [1–3]. Endorsed in January 2015 by the African Heads of State and Government at the 24th Ordinary Session of the Assembly of the African Union, Agenda 2063 is Africa's strategic development program that spells out the continent's economic and social aspirations over the long term [4]. Rooted in Pan-African principles of African solidarity, Agenda 2063 is an endogenous transformation plan that seeks to strategically and competitively reposition the African continent to ensure poverty eradication and equitable people-centric socio-economic and technological transformation. The continent's strategic objectives are broadly encapsulated into seven strategic aspirations that basically articulate the desired state of the African continent. In a nutshell, the aspirations endeavour to spell out Africa's vision regarding inclusive growth and sustainable development, continental integration and Pan-Africanism, good governance and democracy, peace and security, cultural identity and heritage, people-driven development, and the need to be a competitive and strong global player and partner [4–6]. Some of the strategic impact areas identified in the Agenda 2063 implementation plan include wealth creation, shared prosperity, sustainable environment, and transformative capacities [7]. Some of the indicators for wealth creation are access to electricity (% of the population), access to clean water, annual gross domestic product (GDP) growth rate, and industrial development index in Africa. The indicators for sustainable environment include the annual change in forest area (%), the sustainable nitrogen management index, terrestrial sites of biodiversity importance that are completely protected, and freshwater withdrawal as a % of total renewable water resources [6,7].

With a population of approximately 1.225 billion that is highly diverse from 54 countries, Africa is considered to possess the second biggest population in the world. Africa's population is estimated to be growing by at least 2.6% per annum and is estimated to reach 2.5 billion by 2050 [8]. Since 1995, Africa has experienced considerable economic development; in fact, it is the second-fastest developing area in the world. Africa's economic performance is partially attributable to its huge landmass and ocean territory that hosts rich natural resources such as minerals, oil, natural gas, timber, and extensive swaths of rich agricultural lands. Considering Africa's huge landmass and diverse geography, the accomplishment of the Agenda 2063 goals requires systematic spatial planning and data collection, monitoring, evaluation, and consolidation. The aspirations, goals, priority areas, and targets articulated in the Agenda 2063 strategic blueprints are inherently geographic in nature as they all occur in space (location) and time. To this end, geographical information science (GISc), particularly geographic information systems (GIS), earth observation, and global navigation satellite systems can play a fundamental role in supporting the attainment of the Agenda 2063 targets through data collection, storage, integration, evaluation, and reporting.

Since the dawn of human civilization, geographical information tools such as maps and compasses have played a fundamental role in guiding mankind's developmental agenda. Given that location plays a critical role in spatial and public infrastructure development planning, maps have played an integral role in territorial planning, cooperation, navigation, and conflict resolution for many centuries. Historically, maps were also used extensively in the discovery of new territories and natural resources. In the last few decades, the rapid and simultaneous technological developments in computer science, satellite communication, navigation, and earth observation systems have widened the scope of the usage of GISc in many areas of sustainable development [9,10]. For many decades now, many countries have been exploiting GIS for regional and spatial planning [11,12], although in a siloed manner. It is widely acknowledged that location plays a central role in integrating information about the society, the economy, and the environment [13,14]. In addition, cross-border challenges such as climate change, disasters, peace and security, and quality of the environment can only be solved through coordinated global and regional efforts [2,15,16]. In recognition of the power geographical locations and geospatial information play in development planning, the United Nations Statistical Commission endorsed a work program in 2013 to develop a statistical spatial framework that would become an international standard for the integration of statistical and geospatial information [17]. This endorsement comes following the pioneering work done by the UN Global Geospatial Information Management (UN-GGIM) since 2011 to create a framework and define protocols to integrate statistics and geoinformation data processes [18]. The critical role played by timely and consistent geographic information in supporting sustainable development is widely acknowledged by the Group on Earth Observations (GEO) and the Committee of Earth Observation Satellites (CEOS) through their dedicated efforts in coordinating earth observation initiatives globally and advocating for the use of open and free satellite data. The role played by earth observation technology in supporting sustainable development goals is well described in many publications [19–21].

The practical implementation of the African Union (AU) Agenda 2063 blueprint is squarely contingent on the efficient and effective execution of this ambitious and transformative strategic plan. Notably, the strength of any strategic plan lies in its execution capability. Monitoring and evaluation systems play a critical role in collecting, recording, storing, integrating, and evaluating and tracking performance information in the implementation of longer-term strategic plans. Monitoring is the continuous assessment of a plan in relation to the set implementation schedule that is used to provide continuous feedback to management on the implementation progress of the set programme to enable managers to take timely decisions based on identifying successes and constraints. The evaluation uses the data and information provided by a monitoring system to facilitate the analysis of trends and impacts of the programme. Monitoring and evaluation systems are not a new concept, many governments and international institutions have been using them for many years. The practice and culture of monitoring and evaluation are well embedded in organisations such as the United Nations, the World Bank Group and the Organisation for Economic Co-operation and Development (OECD). The United Nations Development Programme (UNDP), for instance, published a handbook to guide the planning, monitoring, and evaluation for development results [22]. The 2009 version of the UNDP handbook [22] integrates planning, monitoring, and evaluation into a single guidebook aimed at inculcating the application of a results-based approach in programming and performance management in order to support the UN development agenda, the Millennium Development Goals (MDGs), at that time. It provides a robust Results Based Management (RBM) results chain that shows the process flow from the inputs, activities, outputs, outcomes, and impacts and the linkage between resources and results. The RBM results chain effectively encapsulates the planning and implementation processes and addresses pertinent strategic planning questions related to execution and impacts. The usage of the GIS as a monitoring and evaluation tool has gained traction in the last few decades due to its ability to support the collection, integration, storage, analysis, output, and distribution of location-based data. While the use of GIS in monitoring and evaluation adds significant value to the current methodologies, most GIS implementations remain institutional without any capability to serve users across institutions, national boundaries, and diverse geographic locations.

The advent of web-based GIS provides a powerful online platform to collect, integrate, discover, use, and share geospatial data, information, and services related to sustainable development [23]. Web-based GIS systems are extending the gamut of geospatial data uses and are helping in improving the update and distribution of spatial data. Cooperative spatial data entry, updates, and management across administrative and political jurisdictions are increasingly becoming feasible using internet-based depots. Geoportals provide a powerful internet-based platform for finding and accessing geospatial information and related geographic services such as sharing, displaying, editing, analysing, and publishing geographic information online [24]. The importance of geoportals as part of spatial data infrastructure is emphasized by Maguire [23]. To support and optimize the widespread application geographic information, most institutions are adopting the use of geoportals to facilitate the discovery, dissemination, and sharing of geospatial information [11,25–29]. The proliferation of geoportals is enabling quick and direct access to geographic data sets of various formats such as satellite imagery, land use, and land cover data sets, and vector layers related to administrative boundaries, transport network layers, and demographic datasets. The geo-visualization capability provided by some deputies enables interactive techniques for visualizing geospatial metadata and information. The growing importance of geoportals is succinctly captured by Crompvoets [30] who sees geoportals as a one-stop-shop for the discovery and assessment of geo-information products and services with significant cost savings. This point is reiterated by Vockner [24] who highlights that geoportals are the main gateways to find, evaluate, and use geographic information. The design and development of most geoportals are largely influenced by the concept of service-oriented architectures (SOA) and the principles and standards prescribed in the Open Geospatial Consortium (OGC), particularly the catalogue service web (CSW) norms [24]. Some of the geoportals that have been successfully implemented to promote and govern the sharing and reuse of geospatial data include the Infrastructure for Spatial Information in the European Community (INSPIRE) Geoportal-European Community Infrastructure for Spatial Information, GeoPortal. Bund for Germany, Geoportail in France, GeoNorge in Norway, and IDEE in Spain [28].

In a recent review of international and national geoportals, an inventory by Jiang [31] showed that most of the geoportals were developed with the prime objective of serving as an electronic infrastructure for e-government and as dissemination platforms for open government data. Some mature national geoportals include Canadas' Open Data, France's geoportal, India's geoportal, United States Geospatial One-Stop, US's geoportal, and China's geoportal. Largely, these portals aim to provide a web platform to enable the sharing of national datasets to increase transparency and accountability in government. Other popular geoportals such as INSPIRE are targeted towards the provision of geospatial datasets to member states of the European Union. Space agencies, surveying, and mapping bureaus are also at the forefront of providing geoportals for the dissemination of satellite data, aerial photography, maps, and other digital geospatial datasets [32]. Popular examples include the European Space Agency (ESA), Copernicus Open Access Hub, and NASA Earth data portal.

Lately, several geoportals have also been launched in Africa. In our review of African geoportals, we observed that most geoportals serve as platforms for geospatial data dissemination for specific projects, examples include the MESA SADC Geoportal, National OCIMS Geoportal, and the WASCAL Geoportal. A similar observation was made for continental, regional, and national geoportals such as the RCMRD Geoportal, Africa Geoportal, and GeoCameron. From our assessment, the current geoportals are not specifically aimed at tracking Agenda 2063 development indicators but serve as online geospatial data distribution platforms. In many cases, the geoportals also lack the basic analytical capability and spatial analysis functionality. While several geoportals already exist internationally as recently reviewed by Jiang [31], currently none of them are targeted towards the monitoring and evaluation needs of Agenda 2063 on a continental scale. Similarly, an inventory of geoportals done by Jiang [31] clearly shows that African countries lack a common online platform for uploading, access, visualizing, storing, analysing, and disseminating development information related to Agenda 2063. To bridge the gaps, there is a need to provide a geoportal dedicated to the monitoring of Agenda 2063 development indicators that has analytical and spatial analysis capability.

To achieve its strategic goals, the African Union (AU) seeks to strengthen its capacity to collect, analyse, and evaluate data to support the implementation of its developmental agenda [4–6]. Acute shortage of data to support the monitoring, evaluation, and comparison of developmental programmes has been stressed as one of the problems facing Africa. AU affirms that the implementation of the Agenda 2063 strategic plan is knowledge-driven and requires both qualitative and quantitative data to support monitoring and evaluation. In the same vein, AU recognises the need for monitoring and evaluation systems at country, regional, economic, community, and continental level to track the implementation of the set targets. The plan also emphasizes the need for the establishment of a common platform to facilitate the preparation of integrated monitoring and evaluation reports. Furthermore, the need to generate reliable, timely, and disaggregated data using appropriate systems for accurate reporting has been highlighted as critical to supporting evidence-based planning.

Recalling that Agenda 2063 aims to achieve socio-economic and political transformation through the optimum use of natural resources, the use of geographic information becomes instructive. Sustainable development is an inherently complex process due to the complex interlinking of information domains that require consideration. The usage of geographic datasets and information systems for planning and decision support permeates across the entire spectrum of sustainable natural resource management. The fusion of web mapping services, spatial visualization, and analysis have considerable potential to support the monitoring and evaluation of sustainable development goals. The South Africa National Space Agency (SANSA) in partnership with New Partnership for Africa's Development-African Union Development Agency (NEPAD-AUDA), the implementing arm of the African Union, has developed a live prototype internet-based geoportal to facilitate data collection, management, integration, analysis, and visualization of Agenda 2063 development indicators. The geoportal is aimed at supporting the planning, implementation, and monitoring of Agenda 2063 goals at continental, regional, and national levels. In this paper, we aim to describe the implementation, structural design, and the functionality of the pilot Agenda 2063 geoportal. Based on the monitoring and evaluation requirements from NEPAD-AUDA, we specifically aim to demonstrate the ability of the geoportal to uploading, accessing, visualizing, storing, analysing, and comparing development information related to the wealth creation, share prosperity, and transformative capacity impact indicators. Selected wealth creation indicators encompassed manufacturing value addition, access to electricity, and improved water sources. Shared prosperity indicators included incidences of tuberculosis, the prevalence of Human immunodeficiency virus (HIV), unemployment with intermediate education, the mortality rate for children under five, and incidences of malaria and gross national income (GNI) per capita growth. Lastly, tax revenue was used as a transformative capacities indicator. Typical users for the geoportal include planning officers from African Union member states, regional economic communities, international development institutions, development aid partners, and funders, NEPAD-AUDA, African Union Commission (AUC), and the public.

2. Data

The success of a geoportal is partly dependent on the availability of the relevant data accessible on the web portal and the frequency at which the data is updated. Several statistical, vector, and raster datasets were sourced from different data providers and uploaded into the geoportal. We obtained statistical socio-economic data from the World Bank, NEPAD-AUDA, and Statistics South Africa to populate the database for this geoportal. The statistical data acquired related to the shared prosperity indicators such as incidence of tuberculosis, the prevalence of HIV, unemployment with intermediate education, the mortality rate for children under five, incidences of malaria, and GNI per capita growth. Data on wealth creation indicators included manufacturing value addition, access to electricity, and improved water sources. Tax revenue data were used as an indicator for transformative capacities indicator. Most of these socio-economic data is updated at least yearly by most countries. Raster and vector data layers were added to provide a spatial and environmental perspective of the situation in various countries. Raster layers consisted of gridded data such as Landsat 8 satellite data acquired from SANSA. The vector datasets layers are comprised of points, line and polygon layers for roads, electricity power lines, administrative boundaries, health service facilities, educational facilities, disaster vulnerability atlas, telecommunication lines, geological maps, rivers, water bodies, and an open street map layer. The vector and raster data were projected to the World Geodetic System (WGS) 84 datum geographic coordinate system. This data are comprised of spatial and non-spatial data. Spatial data included the vector and raster datasets that can be georeferenced to a location on the earth's surface while non-spatial data does not refer to a specific location. While non-spatial data are normally delivered in spreadsheet formats, raster data are supplied in various formats that include GeoTiff, Tiff, JPG, NetCDF, and PNG while vector files are delivered as ESRI shapefiles and other standard textural formats such as KML, GML, GeoJSON, GeoHash, and WKT. The storage, retrieval, visualization, rendering, conversion, and manipulation of vector and raster spatial data and their formats require relational database schemas with GIS functionality as outlined in the architectural design.

3. Geoportal Implementation Framework

In this section, we concisely describe the main processes involved in the development of the Agenda 2063 Geoportal. Concurrent engineering was used to fast-track the development of the pilot geoportal. The prototype geoportal was designed to demonstrate the concept of integrated planning, monitoring, and evaluation in a practical and tangible manner. The functionality of a geoportal can be assessed in terms of its functionality for end-users, geoportal management, geoportal data security, geoportal interoperability, and interface customization [33]. At the pilot phase, we prioritized the functionality of the geoportal in terms of its functionality to end-users, geoportal management, and geoportal data security. End-user functionality would consider aspects such as the discovery of geospatial data resources, a preview of geospatial datasets, and registration of portal users. To achieve this, end-user engagement was concurrently done with portal development.

3.1. Architectural Design

The architectural design of the Agenda 2063 Geoportal is based on three application layers as illustrated in Figure 1. PostgreSQL[a] client/server, conceptual model (refer to the PostgreSQL 11.3 Documentation: The PostgreSQL Global Development Group, 2019). An overview of the architectural design of the Agenda 2063 Geoportal is shown in Figure 1 below.

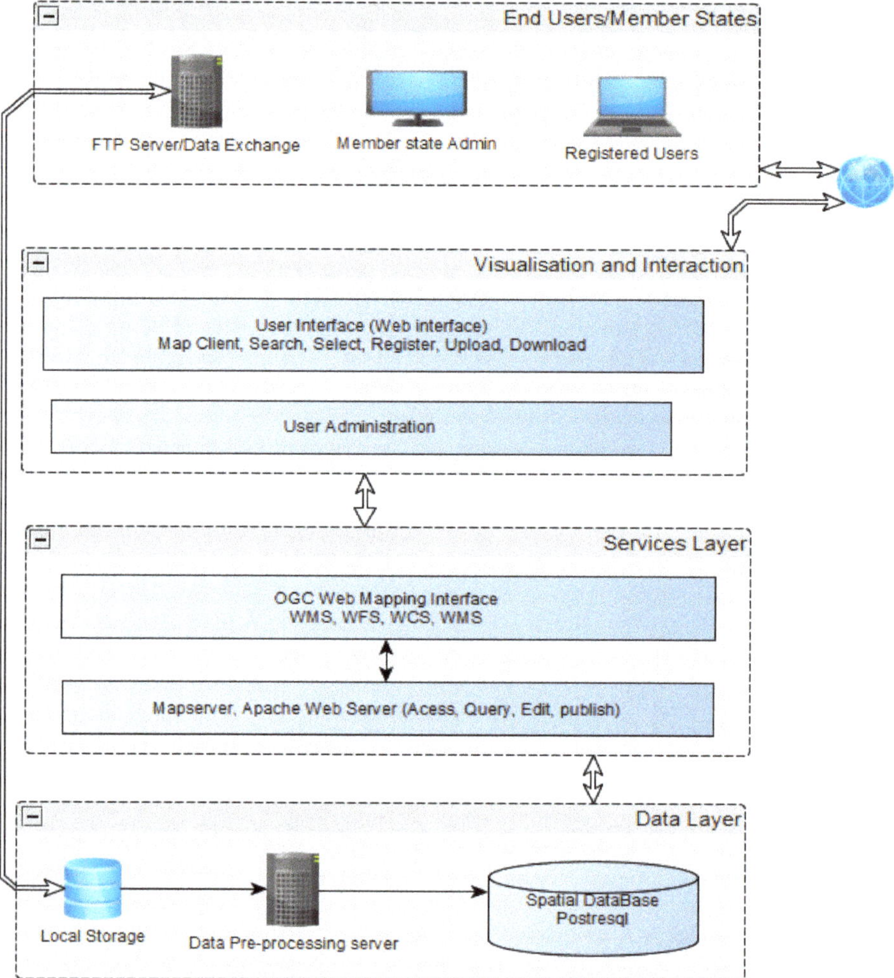

Figure 1. Application layers for the Agenda 2063 Geoportal. Three application layers were implemented for the geoportal.

The first application layer is the database layer that stores and serves all Sustainable Development Goals (SDGs) dataset sourced from different countries. PostgreSQL database with PostGIS extension was selected in the implementation of the geoportal due to its ability to deal with both spatial and non-spatial data types. This then ensured that any data types from 54 African countries can be stored.

PostgreSQL has a proven capacity to deal with a range of large geographical datasets that include vector, raster, and numerical data. PostgreSQL is an open-source object-relational database management system (ORDBMS) that uses a client/server model. In addition to being open-source, some of the attractive features offered by PostgreSQL include its ability to handle complex queries, multi-version concurrency control, triggers, update views, transactional integrity, security enforcement, and its capability to handle foreign key. The backend server-side processes are responsible for managing the database files by accepting connections from client applications and executing the requested database actions using a dataset server program known as PostgreSQL. Some of the basic functions for the server-side include data storage and transformation.

The second layer consists of a data services layer that allows for access, processing, and sharing of data stored in the database. This layer allows for the publishing of interactive maps and spatial data to the web for consumption by various web mapping applications with map visualization capabilities. MapServer, an open-source platform, was used to publish this data is an application-independent manner by using OGC standards thus achieving data interoperability that will allow the sharing of this data with other organisations.

The third layer is the visualisation layer which serves two main purposes namely, the display of data served from the database for the visualisation and the capture of data queries from the users and pass it on the database via the services layer. The results of these queries are also viewed from the visualisation layer. Visualisation is achieved via a web user interface that provides user administration functionalities.

3.2. Tools and Resources

The portal was developed using an ensemble of cross-platform programming and scripting languages that can be used on various operating platforms such as Windows, Linux, and UNIX. The portal's front-end user interface was developed using ExtJS[c] version 6 and GeoExt[d] version 2. ExtJS is a cross-platform web application JavaScript framework and interactive application. GeoExt is also a JavaScript framework cross-platform application that adds Open Layers[e] and GIS functionalities into ExtJS. Agenda 2063 Geoportal uses PHP[f] (Hypertext Pre-processor) open-source, cross-platform language as a server-side language for database querying, deleting, updating, and adding new data. The portal also uses MapServer[g], an open-source, cross-platform application to publish the maps. MapServer uses PostGIS[h] and GeoExt to display the published interactive maps. Figure 2 below provides a concise depiction of the interactions between the front-end and the backend of the Geoportal.

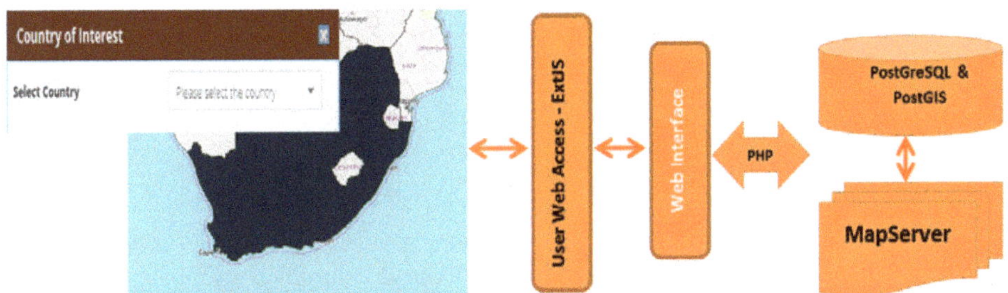

Figure 2. Example of a user query pipeline. Country selection examples from a user click query.

3.3. Implementation of the Backend: Databases

The backend PostgreSQL database management system was spatially enabled using the PostGIS extension. PostGIS, through its generic index structure (GIST), embeds the spatial functionality to enable PostgreSQL to efficiently handle spatial data, indexes, GIS algorithms, and geographical analysis functions. The integration of PostGIS was done through the data connection method aimed at upgrading the PostgreSQL object-relational database into a backend spatial database with GIS capability. Moreover, PostGIS adds indexing, transaction, and concurrency functionality to PostgreSQL. PostgreSQL is a robust object-orientated relational database management system with a proven ability to work in high volume environments. It is ACID (atomicity, consistency, isolation, and durability) compliant. MapServer was integrated to support the display of dynamic geospatial maps over the internet. It supports the display and querying of raster and vector datasets from the database. Aside from querying capabilities, the integration of MapServer was aimed at enabling on the fly projections, spatial analysis and editing, high-quality rendering, cross-platform compatibility, and data interoperability. It has client and backend functionality, enabling the implementation of a hybrid architectural design that is known to improve system efficiency and online geospatial analysis.

3.4. Implementation of the Web Front-End: Geospatial Services

The Agenda 2063 Geoportal aims to provide a wide range of client geospatial web services to African countries. Key among them is the ability to discover, visualize, access, retrieve, compare, and download spatial data sets related to agenda 2063 goals. To build the front-end of the geoportal we integrated map service, feature service, and image service functionality into the web interface. The processes for data access and retrieval are depicted in Figure 2 above. As depicted in the flow chart above, various operations are supported by the web application front end. A selection operation for a country of interest is illustrated where a user click is used to query the database and return a country of interest. Similar operations are supported for selecting and displaying country indicators and statistical information that is then analysed and compared by various charting options provided by the front end.

Three levels of user access were defined in the design of the geoportal. The general user access level is aimed at the general public. At this level, users can use some of the most basic functionality of the geoportal which includes visualization of the geospatial datasets available on the portal. The general public is required to register and login to access additional geospatial services, such as charting for comparing development data from countries and regions. To protect the integrity of some sensitive datasets, access rights are validated using a filter once a user has logged in. User rights are administered by the national statistical offices. The data administrators are responsible for data verification, data uploading, and data integrity and data access in their own countries. A system administrator is assigned responsibility for managing the computer systems, web pages, databases, security threats, and the environment where the system is operating.

Since MapServer was utilized in achieving data interoperability, through OGC data sharing standards, datasets can be served to third-party visualization platforms such as QGIS and ArcMap by WMSs. When the user requests data from the portal, the portal first validates the user access rights and once permission has been granted the data is retrieved from the host server database and sent back to the user. The retrieval process is in line with the Open Geospatial Consortium[i] standard for web mapping as shown in Figure 3 below.

3.5. Geospatial Datasets and Development Indicators

Agenda 2063 has 20 strategic goals that crosscut the seven aspirations described earlier. For demonstration purposes, we selected a few economic and health development indicators to showcase the functionality of the geoportal. Development indicators used in the development of the geoportal include GNI per capita growth (annual %), mortality rate for children under five years per 1000 live births, access to electricity (% of population), unemployment with intermediate education, improved water sources, tax revenue (% of GDP), prevalence of HIV, total (% of population), manufacturing, value added (% of GDP), incidences of tuberculosis (per 1000 people) and incidences of malaria. Our aim was to use public domain datasets. We integrated a number of geospatial datasets to supplement visual analysis and these include vector and raster data layers. The vector and raster layers include base layers of member states, transport networks, geology, population, industry, information and communication networks, human settlements, disaster, sustainable land and water management, energy, health services, education, and satellite imagery.

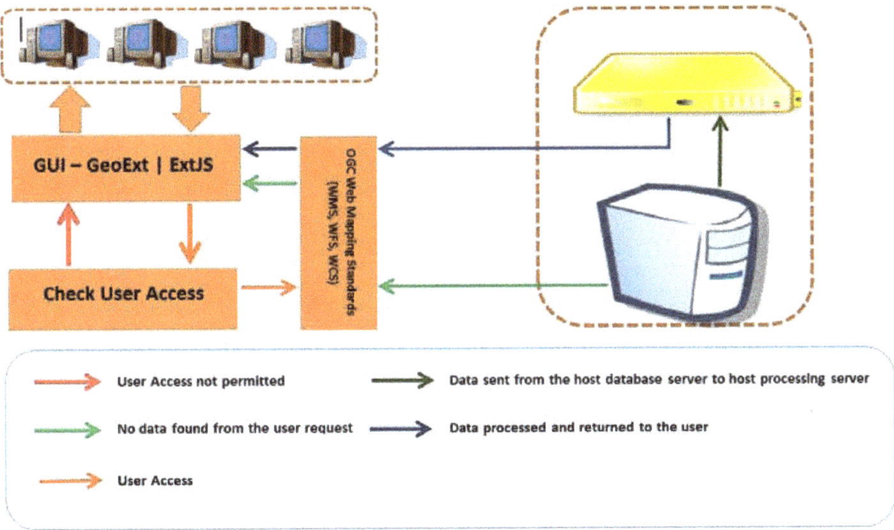

Figure 3. Data access restrictions. Data interoperability is achieved using Open Geospatial Consortium (OGC) standards. However, access to sensitive datasets is restricted at the user level.

3.6. Awareness and User Requirements Collection Seminars

We conducted seminars with officers involved in planning, monitoring, and evaluation to raise awareness on the value of the Agenda 2063 geoportal as a tool to support integrated spatial planning, monitoring, evaluation, and reporting. A second objective of the seminars to collect user requirements and understand the current monitoring systems and evaluation systems currently in use. The participants were drawn from Niger, Southern African Development Community (SADC), East African Community (EAC), the Intergovernmental Authority on Development (IGAD), NEPAD-AUDA, and the AUC. The seminars we conducted in Durban and Johannesburg in South Africa and Niamey in Niger. Awareness sessions we also conducted by NEPAD-AUDA in various African countries.

3.7. Hardware Infrastructure

The pilot geoportal application was deployed on a development server with 300 GB storage capacity, 32 GB memory, 2 processor sockets, each with 16 cores. A storage server with over 30TB of storage space was mounted on the deployed server for storing the large continental base layer datasets.

4. Demonstration

In this section, we aim to present the functionality of the prototype geoportal. We will showcase its front, user access, and data upload interfaces, development indicators, and charting capabilities. Furthermore, we will demonstrate the ability of the geoportal to display development trends such as access to electricity, manufacturing value addition, GNI per capita growth, and tax revenue information. We also highlight the geoportal's capability to tabulate inter-country comparisons of specified development indicators. The visual and integration functions are depicted by applying map overlay operations using the human settlements layer and SPOT 6 satellite imagery.

The Agenda 2063 Geoportal shown in Figure 4 in this project demonstrates that geoportals are a powerful web-based interactive platform to upload, access, explore, visualize, analyse, and disseminate geospatial data related to the sustainable development of the African continent. Although in the pilot phase, the Geoportal demonstrates the primary functionality of geoportals in terms of its capability to discover, analyse, share, and download geospatial datasets as evident on the following web link: http://agenda63sdgs.sansa.org.za/nepad/.

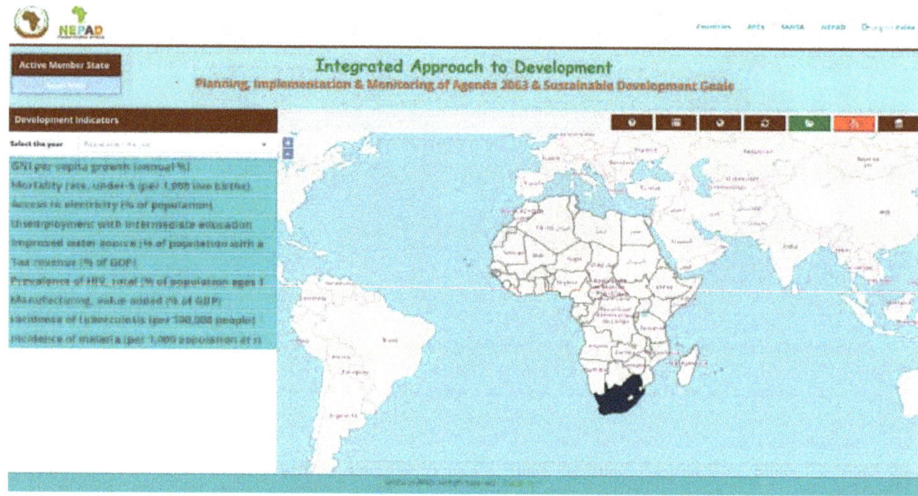

Figure 4. Front end interface for interacting with the geoportal.

In this case, we used a number a geo-datasets such as statistical development indicators, demographic statistics, satellite imagery, vector data sets for transport networks, and administrative boundaries, hazard layers, land use and land cover base layers, and digital elevation models as shown by the product layers in Figure 5.

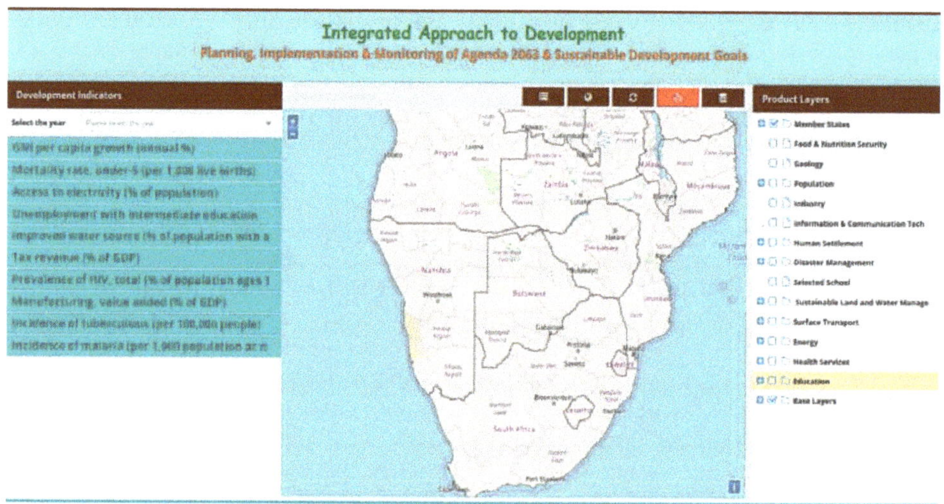

Figure 5. Base layers can be viewed on the maps and overlaid for visual and quantitative analysis.

The interactive nature of the Geoportal allows the overlay of vector and raster datasets and to make comparisons between countries, regions, and provinces aided by graphical analysis. The browsing and viewing capability are aimed at the general public and more advanced functionality such as graphical comparisons is designed for authorized users of the geoportal. In order to protect the integrity and confidentiality of the datasets, three levels are users are designated. Access to information and more advanced functionality of the geoportal is granted through user registration and password security as shown in Figure 6 below.

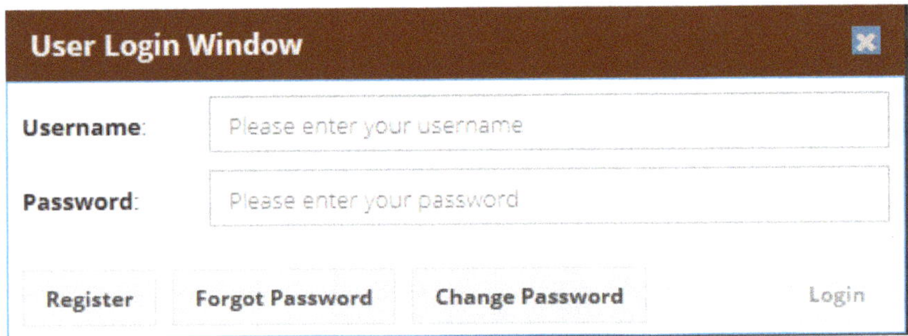

Figure 6. User rights administration.

The geoportal has the capability to allow authorized officials from national statistical offices to independently upload official datasets as shown in Figure 7 below.

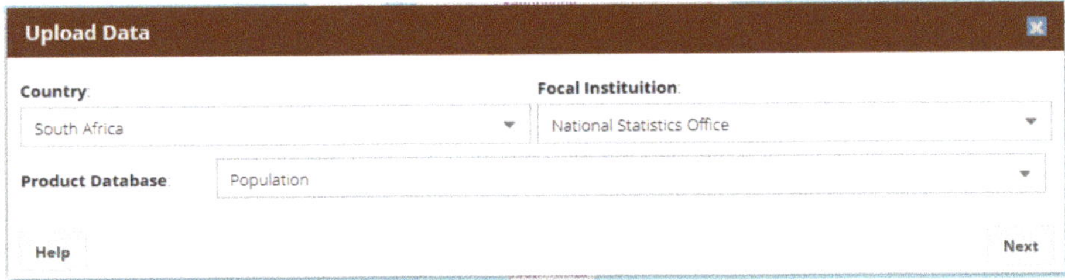

Figure 7. Data upload capabilities for data updating and sharing.

In this geoportal, we selected several development indicators related to the strategic impact areas of shared prosperity, wealth creation, and transformative capacities. The snapshot in Figure 8 below illustrates the results of our implementation. We show for each country, the high-level strategic impact, the indicator, and the Agenda 2063 target.

Country	Strategic Impact	Indicator	Agenda 2063 Target
South Africa	Shared Prosperity	Incidence of tuberculosis (per 100,000 people)	Reduce the 2013 incidence TB by at least 80%
South Africa	Shared Prosperity	Prevalence of HIV, total (% of population ages 15-49)	Reduce the 2013 incidence of HIV/AIDs by at least 80%
South Africa	Shared Prosperity	Unemployment with intermediate education	
South Africa	Shared Prosperity	Mortality rate, under-5 (per 1,000 live births)	Reduce 2013 child mortality rates by at least 50%
South Africa	Shared Prosperity	Incidence of malaria (per 1,000 population at risk)	Reduce the 2013 incidence of Malaria by at least 80%
South Africa	Transformative Capacities	Tax revenue (% of GDP)	Tax and non-tax revenue of all levels of government should cover at least 75% of current and developme...
South Africa	Shared Prosperity	GNI per capita growth (annual %)	Increase 2013 per capita income by at least 30%
South Africa	Wealth Creation	Access to electricity (% of population)	Increase access and use of electricity and internet by at least 50% of the 2013 levels
South Africa	Wealth Creation	Improved water source (% of population with access)	Reduce 2013 level of proportion of the population without access to safe drinking water by 95%
South Africa	Wealth Creation	Manufacturing, value added (% of GDP)	Real value of manufacturing in GDP is 50% more than the 2013 level

Figure 8. High-level strategic impacts, indicators, and the Agenda 2063 targets.

The charting capability of the geoportal is demonstrated in Figures 9 and 10 below. We show the ability of the geoportal to provide graphical plots of the development indicators over time to support trend analysis. In this paper, we used the access to electricity development indicator to showcase the geoportal graphical plotting capabilities.

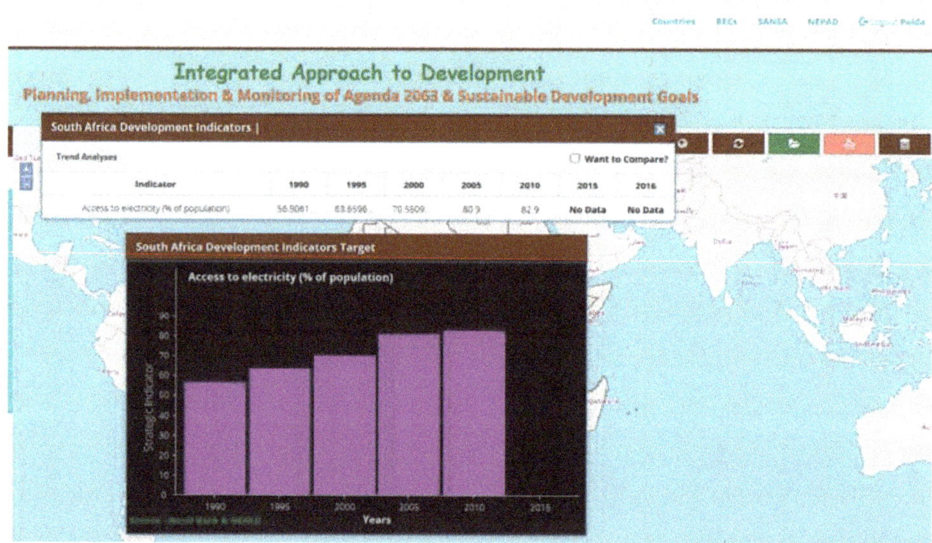

Figure 9. Charting options are provided by the front end for visual analysis.

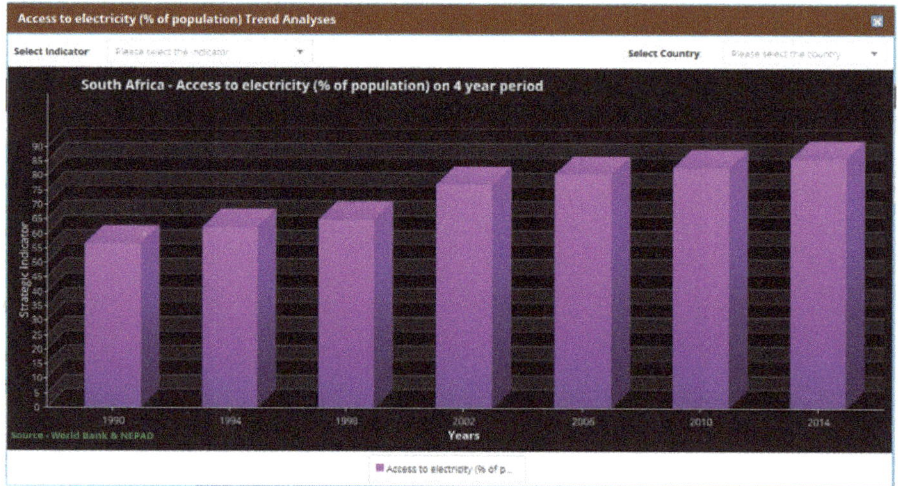

Figure 10. Example of a chart displaying the percentage of the population with access to electricity and how this varies over time.

Comparison of development indicators between different member states and regional economic communities is another essential function offered by the geoportal as shown in Figure 11. This functionality allows users to compare the rates of development among African states and learn from the different developments interventions implemented in countries that show better growth. A comparison of access to electricity shown below for South Africa, Botswana, Zambia, and Zimbabwe.

Member States ↑	Region	Member State...	1990	1992	1994	1996	1998	2000	2002	2004	2006	2008	2010	2012	2014
Algeria	Northern Africa	Botswana	5.807...	10.15...	14.48...	18.78...	23.03...	27.2...	31.3...	35.4...	39.6...	43.1	48.0...	52.2...	56.4...
Angola	Southern Africa	Equatorial Guinea	54.83...	56.01...	57.18...	58.32...	59.40...	60.4...	61.3...	62.3...	63.3...	64.3	65.4	66.4...	67.5...
Benin	Western Africa	South Africa	56.50...	59.37...	62.23...	57.6	64.9	70.5...	77.1	80.9	80.7	81.9	82.9	85.3	86
Botswana	Southern Africa	Zambia	13.9	19.2	16.15...	17.3	17.87...	16.7	17.4	20.3	20.7...	21.5	22	23.1...	27.9

Figure 11. Data comparison for four countries. Access to electricity comparison.

The ability to query a geoportal is critical for many users. Here, we demonstrate this functionality by querying the education test dataset as shown in Figure 12 below.

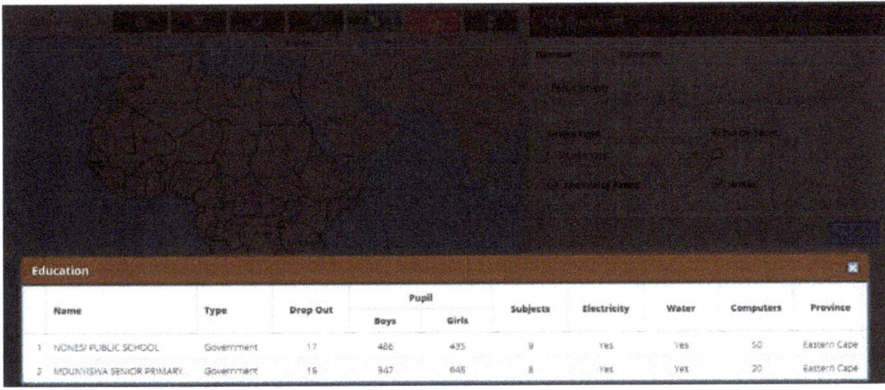

Figure 12. Education information queries displayed by the front end.

Information regarding the strategic impact areas can also be graphically displayed on the geoportal. Figure 13 below illustrates the temporal trends for manufacturing, value-added from 1990 to 2015 and shows the expected target. For manufacturing, the value added is a wealth creation indicator. Similarly, the GNI per capita growth is shown in Figure 14 and mortality rate trends are shown in Figure 15 for the same period. GNI per capita growth and mortality rate for children are the five years, and are both shared prosperity indicators. Tax revenue, a transformative capacity indicator is shown in Figure 16.

The strength of the Agenda 2063 lies in its ability to integrate and visualize the various statistical development indicators with geographical vector and raster datasets. In Figure 17, we show the ability of the geoportal to display human settlements layers over the city of Niamey in Niger.

Satellite images provide a powerful means of visualizing urban and rural landscapes over time. They provide a synoptic view of the environment and provide visual evidence of sustainable development on the ground. Figure 18 below illustrates the capability of the portal to display satellite imagery as visual evidence in analysing sustainable development trends and impacts such as urbanization.

The geoportal is spatially enabled to perform a wide range of spatial analytical functions through the integration of the PostGIS onto the PostgreSQL relational database. Some of the basic functions currently performed by the geoportal include buffer analysis, map overlay, spatial filtering, attribute queries, spatial queries, distance querying, and calculation of areas and length. The attribute and spatial querying functionality are shown in Figures 19 and 20 respectively. Figure 21 illustrates the buffering capability of the geoportal in the selection of schools within a specified distance from a selected school.

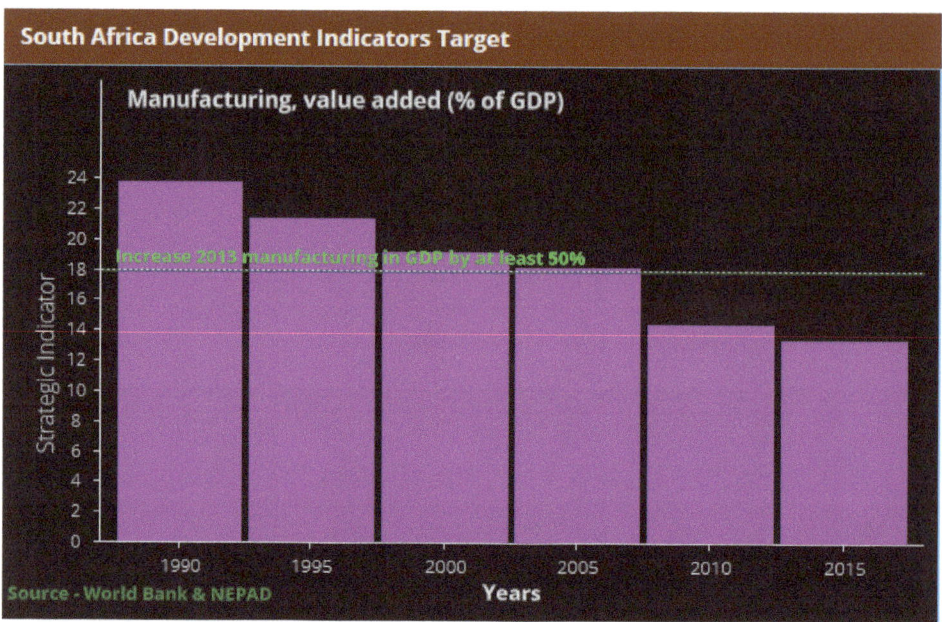

Figure 13. Manufacturing, value-added information displayed by the front end.

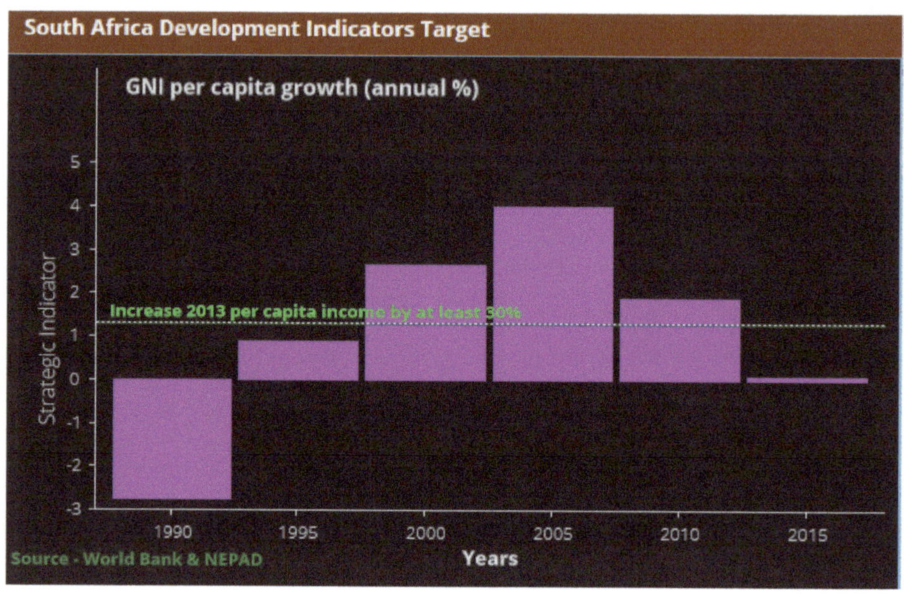

Figure 14. Gross national income (GNI) per capita growth information displayed by the front end.

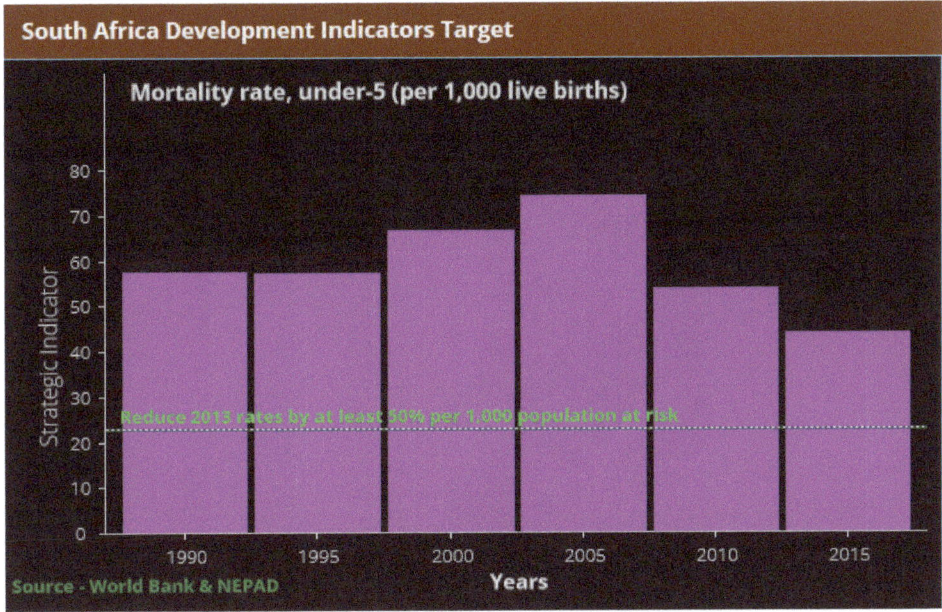

Figure 15. Mortality rate information displayed by the front end.

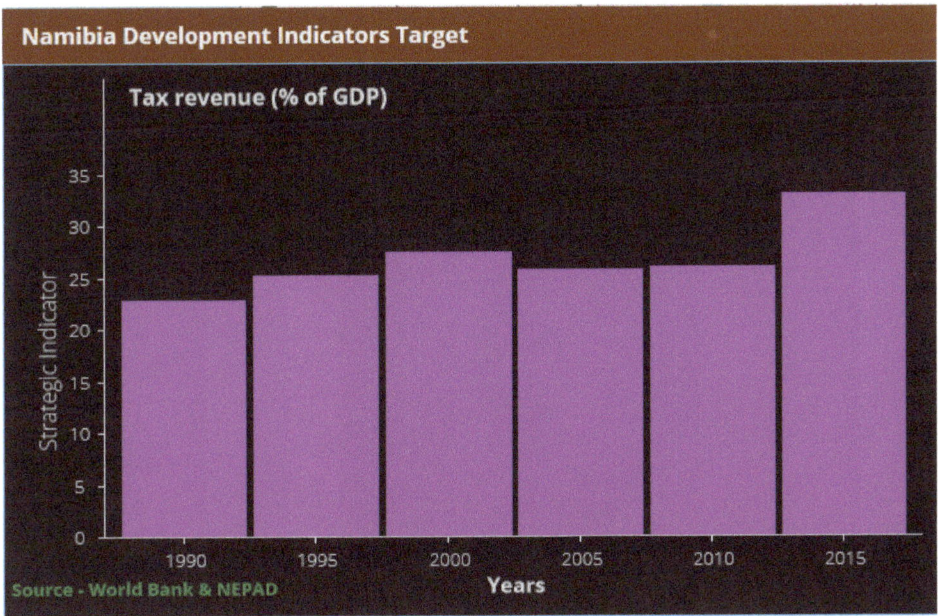

Figure 16. Tax revenue information displayed by the front end.

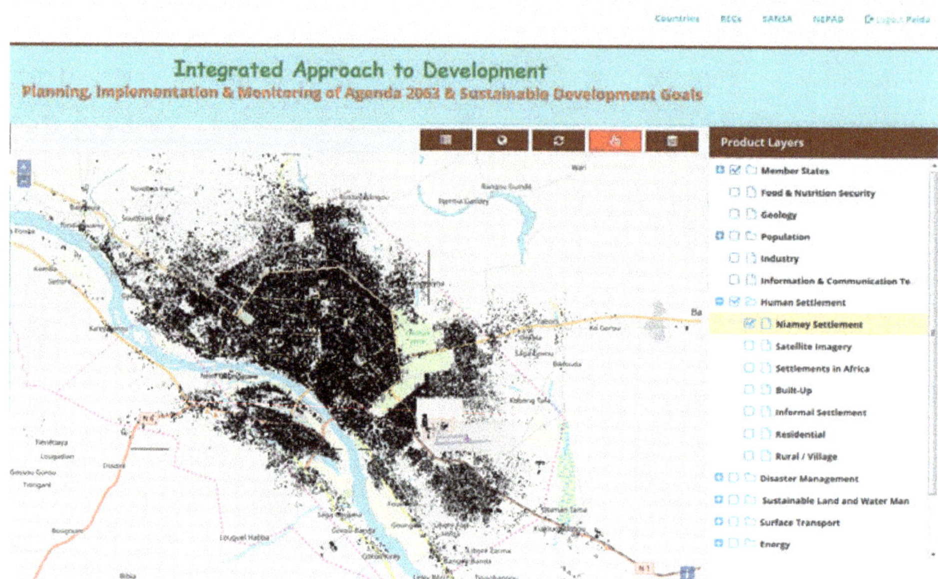

Figure 17. Map overlay operations. Human settlements layer over the city of Niger.

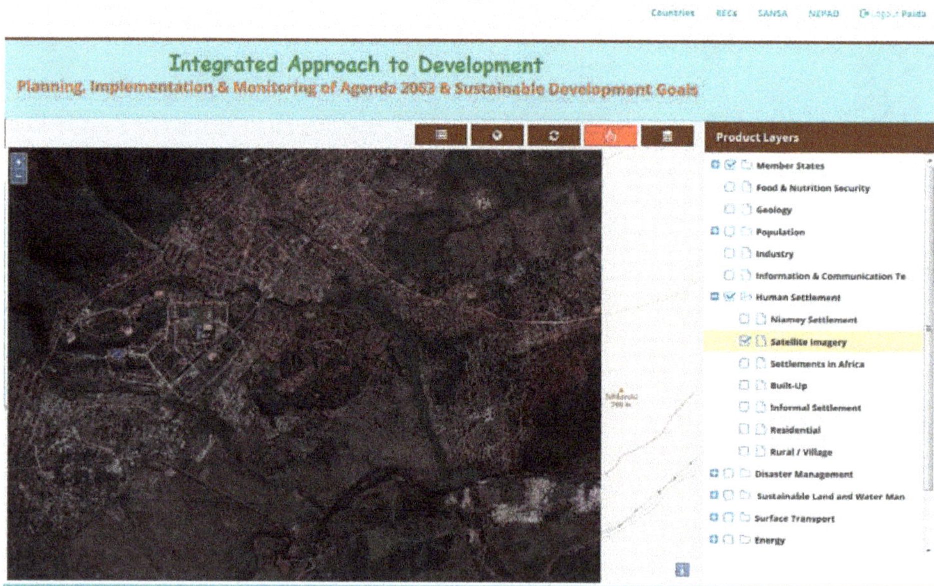

Figure 18. Satellite image data overlays.

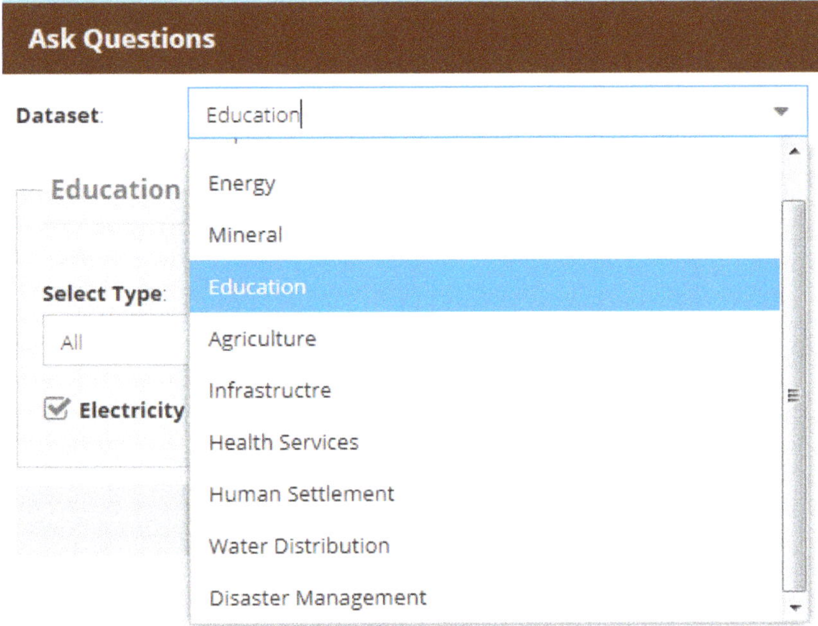

Figure 19. Attribute querying capability.

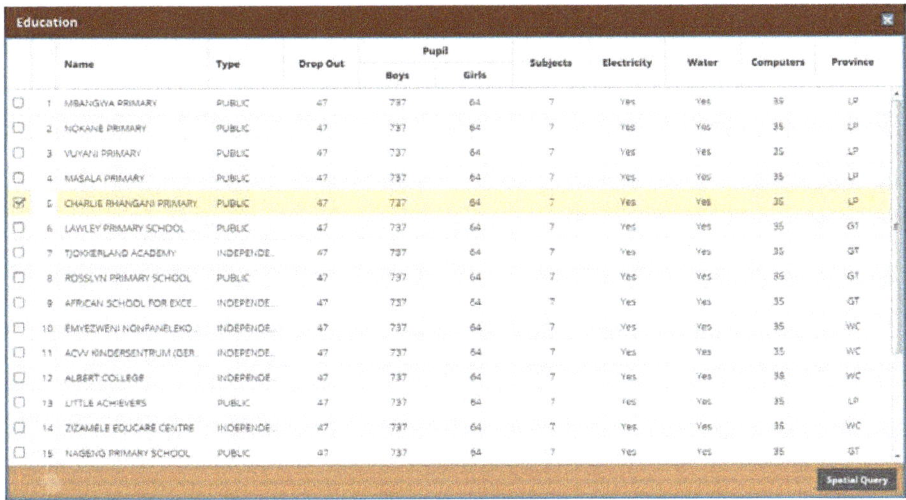

Figure 20. Spatial querying capability.

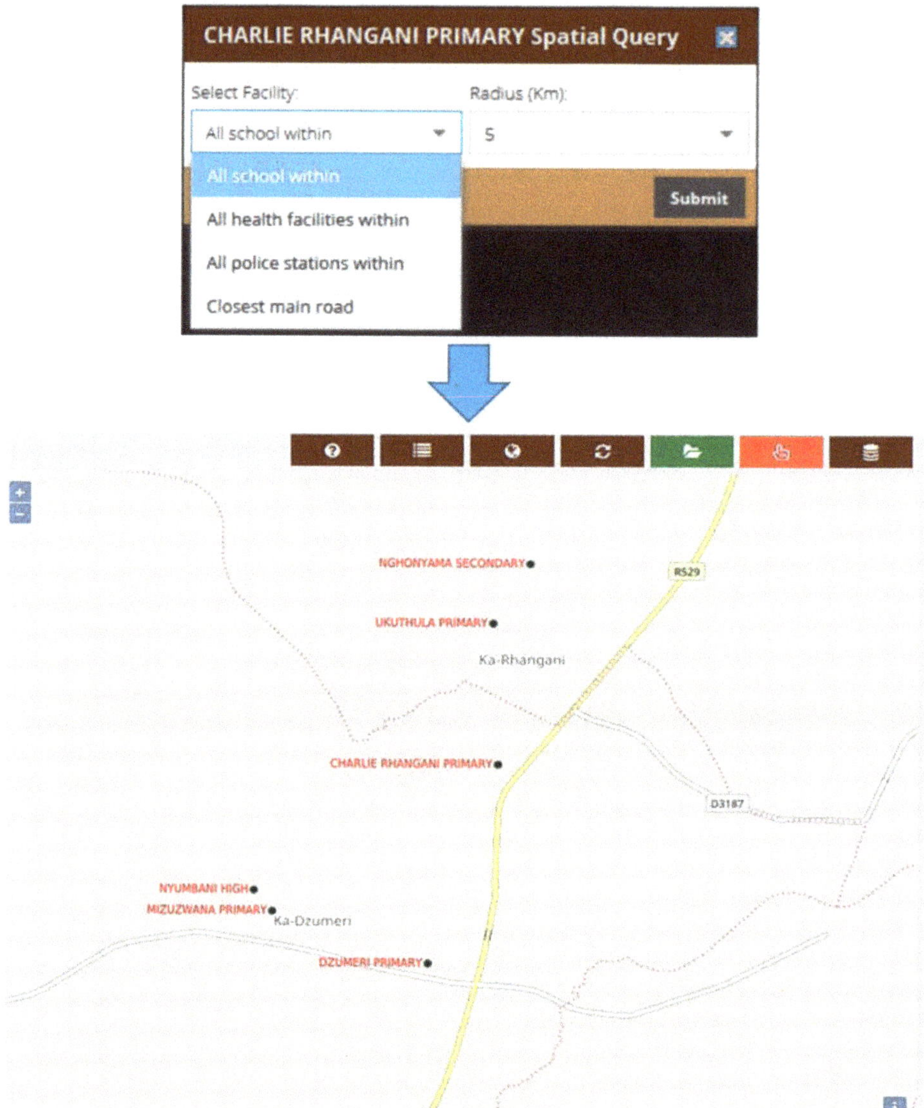

Figure 21. Buffering schools within a 5 km radius of Charlie Rhangani Primary.

5. Feedback from User Seminars

In all the user workshops conducted, the participants supported the need for the piloted Geoportal as it provides an integrated digital platform for the discovery, visualization, analysis, upload, and download of development statistics for Africa. The geoportal was seen as a fit-for-purpose tool that will support the monitoring and evaluation of Agenda 2063 targets. While participants supported the geoportal functionality and the basis for its development, some recommendations were provided. Pertinent issues that emerged from the workshop included issues regarding data policy and custodianship, mandates for Agenda 2063 and SDG reporting in various countries, access rights, maintenance of data integrity, validation of data sources, data provenance and metadata, geoportal security, the synergy between Agenda 2063 targets and the UN SDGs, data collection methods, and data standards. The participants also emphasized the need to provide training to public officials involved in planning, monitoring, and evaluation on how to use the geoportal to access and discover information for analysis and reporting. Participants also recognized the need for a federated system when upscaling the prototype geoportal into a functional geoportal. A key message that also emerged from the workshop was the need for linking the geoportal to other existing geoportals and development-oriented systems to facilitate data harvesting, interoperability, and synergy with countries and regional economic communities. These recommendations were accepted and will be implemented when upgrading the prototype geoportal.

6. Discussion and Future Direction

While our prototype geoportal provided the basic functionalities of a geoportal such as data uploading, querying, visualization, graphical functionality, and downloading capability for demonstration purposes, additional functionality is required for an operational geoportal. Future work will be oriented towards enhancing the analytical capability of the geoportal, dynamic visualization, 3-dimensional visualization, metadata cataloguing, provenance tracking, user feedback functionality, and enhancement of the search functionality using grid-enabled search techniques and web crawlers. Katumba [34] provides a valuable overview of advanced search techniques optimized to improve the discovery of geospatial data on the internet. Given the sheer volume and variety of data to be stored on the geoportal, cloudification of the geoportal will be considered. Iosifescu-Enescu [35] indicates that auto-scalable cloud-based architectures using virtual servers and autoscaling have advantages such as rapid elasticity and on-demand self-service. Alternatively, a federation or hybrid system architecture that is capable of connecting to multiple distributed systems will also be considered. Recently, Jiang [31] suggested solutions and recommendations regarding geoportal architectures, services, and state of the technologies to deal with common geoportal problems. Li [36] showcased many visualization techniques based on spatial scale and temporal primitives that have the potential for adoption. The visual information seeking mantra (VISM) technique showcased in their paper is beneficial when exploring data at a global, group, and local scale while classic visualization templates offer an opportunity to present charts and graphs to support monitoring and evaluation.

While we managed to demonstrate the spatial functionality of the geoportal, future work will be dedicated towards enhancing this functionality. Future upgrades will be directed towards the integration of the open-source R statistical software to enhance the statistical and spatial data analytical capability of the geoportal. Advanced spatial analysis is currently feasible through a QGIS software WMS connection. Future developments of the geoportal will also be directed on customizing the geoportal in line with the requirements of the various user segments. The front end interface of the geoportal accessed will be configured by changing language options and providing menu choices and buttons that are specific to countries and regional economic communities. Using interactive graphical customisation, the geoportal will also be adapted to change colour schemes and icons in line with the user needs. Planning officers that use the geoportal frequently for reporting purposes will have access to programmed functionality aimed for optimized reporting. Customization of the geoportal will also include the provision of spatial and statistical analysis tools and data translators designed to support advanced spatial planning and analysis. Besides, database management systems will be customized to enable connectivity and interoperability with other ratified geographically distributed spatial databases.

The operational implementation of the geoportal should conform to international geospatial standards such as the OGC, Spatial Data Infrastructure (SDI), Dublin Core metadata standard, Schema.org standard, and 1SO/TC211 standards. Best practices and guidelines provided by UN Statistics such as the Fundamental Principles of Official Statistics, Shared Guiding Principles for Geospatial Information Management and the Integrated Geospatial Information Framework will also be adopted to ensure data quality. Data integrity shall also be ensured by a peer-review process as set out by in the African Union Implementation Plan.

From the functionality illustrated in this paper and feedback collected from the user seminars, it is evident that the Agenda 2063 Geoportal will play a significant role in providing geoinformation products and services necessary to support evidence-based integrated spatial planning. It is foreseen that with the demonstrated capability, the Agenda 2063 Geoportal has the potential to play an important role assisting the tracking of the sustainable development indicators over time, and will support data-driven decision making across sectors, within continental, regional, national, and sub-national spheres. The geoportal implemented in this project provides a credible mechanism to share credible, timely, and consistent data and information to support sustainable development enabling efficient monitoring and evaluation. Timely information enables timely decision-making.

Future development options would look at the potential of a geographically distributed geoportal. While noting the data storage complexity associated distributed GIS databases over centralized databases, Kumar [37] highlights a number of distributed GIS databases are more efficient in terms speed, reliability, and storage efficiency when sharing large heterogeneous datasets. Distributed geoportals are particularly suitable when users require real-time responses of their queries and when the geoportal has interactive graphics. The deployment of the geoportal onto a cloud platform is another alternative that could be considered in order to achieve high operational efficiency in terms of data storage, upload, and analytical capabilities. Cloud computing is increasingly being used in the geospatial sciences to storage and analyse large datasets such as satellite imagery. Amazon web services (AWS) have already demonstrated the feasibility of hosting such as geoportal on a cloud platform through the hosting of the Africa Regional Data Cube and Open Data Cube Initiatives (https://aws.amazon.com/blogs/publicsector/tag/africa-regional-data-cube/, 2019). The integration of the geoportal with emerging digital platforms for accessing and analysing satellite imagery for assessing environmental conditions over the continent will enhance the capability of the geoportal. Interfacing the geoportal with the Digital Earth Africa data cube platform (http://www.ga.gov.au/digitalearthafrica), for instance, will open up opportunities for long-term time series analysis and change detection of urban areas, surface water resources, croplands, vegetation cover, land degradation, and other natural resources necessary to achieve Africa's development agenda.

7. Conclusions

Our prototypic geoportal implementation demonstrates that a web- geoportal can be used for accessing, integrating, visualizing, analysing, and disseminating multi-datasets to support the monitoring of the Agenda 2063 sustainable development indicators. It provides a single data platform for integrating multiple data sources that can be compared over space and time. Most importantly, the geoportal allows governments across Africa to share their development information in support of Agenda 2063 and enables them to visualize and to compare their data with those of other African countries on a common online platform. Timely information is a critical input in the design, monitoring, and evaluation of multi-sectoral sustainable development policies. The integration of satellite imagery into the Geoportal provides a data visualization functionality that can be used to visually illustrate the current situations with regards to access to basic services such as schools, health care facilities, electricity, urban and infrastructure development, flooding events, agriculture, and surface water resource evaluations. The graphical functionality of the Geoportal allows users to monitor trends between different countries and regional blocks. The Geoportal provides an open and common platform for sharing, monitoring, and evaluation in support of evidence-based planning, reporting, and decision-making. In this paper, we have been able to demonstrate that while most geoportals have a national and project focus, it is feasible to have a continent-wide geoportal that serves a common digital platform for data discovery, visualization, storage, analysis, uploading, and dissemination. This geoportal will assist in providing access to sustainable development data aligned with Agenda 2063 and will enable the user to compare development trends across Africa. It will also allow development funders to quickly access information on the impact of their development programmes in different regions of the continent on a data-rich web platform.

The success of a geoportal is evaluated by its functionality for end users, geoportal management, geoportal data security and geoportal interoperability, and interface customization. We aimed to develop a demonstration geoportal through rapid prototyping and concurrent engineering. Through the demonstration held with the regional economic communities in Africa, the African Union Commission, and selected member states the value of a geoportal for integrated reporting, monitoring, and evaluation were ascertained. The Geoportal was seen as a pragmatic and unified way of consolidating development information to support the monitoring of the Agenda 2063 goals. The user functionality demonstrated by the portal in terms of data visualization, querying, discovery, analysis, charting, and sharing is valuable for integrated planning, monitoring, and evaluation. Access rights are a key issue in information management, in this, we successfully showed the functionality of the monitoring and evaluation geoportal for by providing registration and access rights to three levels of users that are can be authenticated and validated. The uploading of official and credible information was also an important consideration in the development of this geoportal. We achieved this through a validation done after registration on the system, whereby rights to upload relevant information is only granted to approved officials in member states. The Geoportal management is ensured by granting administrator and publisher User-IDs and passwords. Geoportal data security was achieved through user authentication options available in PostgreSQL. To ensure interoperability, the geoportal supports a wide range of metadata standards and a variety of common electronic data communication standards. Given the successful demonstration of this live geoportal prototype, a consolidated user requirement will be drawn up and an operational geoportal will be developed as the next step. More advanced geospatial and statistical analytical functionality will also be added to the geoportal as it evolves into its operational phase, in compliance with the user requirements. Future considerations will also focus on data policies and data security matters. The full implementation of the Agenda 2063 geoportal into operational mode will be critical to support the tracking performance indicators and evidence-based policy and decision making in Africa.

Author Contributions: Paidamwoyo Mhangara conceptualized, wrote, edited, and submitted the paper and led the implementation of the Geoportal. Asanda Lamba wrote the paper and implemented the Geoportal. Willard Mapurisa wrote, edited the paper and assisted in implementing the Geoportal. Naledzani Mudau wrote and edited the paper.

Funding: This research received no external funding.

Acknowledgments: We are highly indebted to Mohamed Abdisalam, Martin Bwalya, and Simon Kisira all from NEPAD-AUDA for their valuable contribution in guiding the development of this geoportal.

Conflicts of Interest: The authors declare no conflict of interest.

References

1. DeGhetto, K.; Gray, J.R.; Kiggundu, M.N. The African Union's Agenda 2063: Aspirations, challenges, and opportunities for management research. *Afr. J. Manag.* **2016**, *2*, 93–116. [CrossRef]
2. Sirisena, P.; Noordeen, F.; Kurukulasuriya, H.; Romesh, T.A.; Fernando, L. Effect of Climatic Factors and Population Density on the Distribution of Dengue in Sri Lanka: A GIS Based Evaluation for Prediction of Outbreaks. *PLoS ONE* **2017**, *12*, e0166806. [CrossRef]
3. Vickers, B. *A Handbook on Regional Integration in Africa: Towards Agenda 2063*; Commonwealth Secretariat; Marlborough House: London, UK, 2017.
4. African Union. *AGENDA 2063: The Africa We Want Framework Document Adopted in September*; African Union Commission: Addis Ababa, Ethiopia, 2015.
5. African Union. *Agenda 2063: First Ten-Year Implementation Plan 2014–2023*; African Union Commission: Addis Ababa, Ethiopia, 2015.

6. African Union. *Others MDGs to Agenda 2063/SDGs Transition Report 2016: 2016 Towards an Integrated and Coherent Approach to Sustainable Development in Africa*; African Union Commission: Addis Ababa, Ethiopia, 2016.

7. NEPAD. *Integrated Territorial Development. Application of Geographic Information System for A63*; Springer: Basel, Switzerland, 2015.

8. United Nations. World Population Prospects 2019. 2019. Available online: https://population.un.org/wpp/ (accessed on 4 September 2019).

9. Man, W.E.; van den Toorn, W.H. Culture and the adoption and use of GIS within organisations. *Int. J. Appl. Earth Obs. Geoinf.* **2002**, *4*, 51–63.

10. Scott, G.; Rajabifard, A. Sustainable development and geospatial information: A strategic framework for integrating a global policy agenda into national geospatial capabilities. *Geo-Spatial Inf. Sci.* **2017**, *20*, 59–76. [CrossRef]

11. Bao, Y.-W.; Yu, M.-X.; Wu, W. Design and Implementation of Database for a webGIS-based Rice Diseases and Pests System. *Procedia Environ. Sci.* **2011**, *10*, 535–540. [CrossRef]

12. Harris, T.M.; Elmes, G.A. The application of GIS in urban and regional planning: A review of the North American experience. *Appl. Geogr.* **1993**, *13*, 9–27. [CrossRef]

13. Jovanovic, V.; Njegus, A. The application of GIS and its components in tourism. *Yugosl. J. Oper. Res.* **2008**, *18*, 261–272. [CrossRef]

14. Sheppard, E. GIS and Society: Towards a Research Agenda. *Cartogr. Geogr. Inf. Syst.* **1995**, *22*, 5–16. [CrossRef]

15. Frolova, N.; Bonnin, J.; Larionov, V.; Ugarov, A. Complexity in seismic risk assessment at different levels with GIS technology application. In *Engineering Geology for Society and Territory*; Springer: Cham, Germany, 2015; pp. 381–385.

16. Gharbia, S.S.; Gill, L.; Johnston, P.; Pilla, F. Multi-GCM ensembles performance for climate projection on a GIS platform. *Model. Earth Syst. Environ.* **2016**, *2*, 102. [CrossRef]

17. Van Halderen, G.; Minchin, S.; Brady, M.; Scott, G. Integrating statistical and geospatial information, cultures and professions: International developments and Australian experience. *Stat. J. IAOS* **2016**, *32*, 1–14. [CrossRef]

18. Arozarena, A.; Villa, G.; Valcárcel, N.; Pérez, B. INTEGRATION OF REMOTELY SENSED DATA INTO GEOSPATIAL REFERENCE INFORMATION DATABASES. UN-GGIM NATIONAL APPROACH. *Int. Arch. Photogramm. Remote Sens Spat. Inf. Sci.* **2016**, *41*.

19. Anderson, K.; Ryan, B.; Sonntag, W.; Kavvada, A.; Friedl, L. Earth observation in service of the 2030 Agenda for Sustainable Development. *Geo-Spat. Inf. Sci.* **2017**, *20*, 77–96. [CrossRef]

20. GEO. *SDG Coordination across GEO Work Programme Activities*; Group on Earth Observation: Geneva, Switzerland, 2017.

21. Masó, J.; Serral, I.; Domingo-Marimon, C.; Zabala, A. Earth observations for sustainable development goals monitoring based on essential variables and driver-pressure-state-impact-response indicators. *Int. J. Digit. Earth* **2019**, 1–19. [CrossRef]

22. Menon, S.; Karl, J.; Wignaraja, K. Handbook on planning, monitoring and evaluating for development results. *UNDP Eval. Off. N. Y.* **2009**.

23. Maguire, D.J.; Longley, P.A. The emergence of geoportals and their role in spatial data infrastructures. *Comput. Environ. Urban Syst.* **2005**, *29*, 3–14. [CrossRef]

24. Vockner, B.; Mittlböck, M. Geo-Enrichment and Semantic Enhancement of Metadata Sets to Augment Discovery in Geoportals. *ISPRS Int. J. Geo-Inf.* **2014**, *3*, 345–367. [CrossRef]

25. Crăciunescu, V.; St, C.; Ovejeanu, I. Developing an open Romanian geoportal using free and open source software. *Geogr. Tech.* **2008**, *3*, 15–20.

26. Foerster, T.; Schäffer, B. A client for distributed geo-processing on the web. In Proceedings of the International Symposium on Web and Wireless Geographical Information Systems, Cardiff, UK, 28–29 November; Springer: Berlin/Heidelberg, Germany, 2007; pp. 252–263.

27. Fritz, S.; McCallum, I.; Schill, C.; Perger, C.; See, L.; Schepaschenko, D.; Van Der Velde, M.; Kraxner, F.; Obersteiner, M. Geo-Wiki: An online platform for improving global land cover. *Environ. Model. Softw.* **2012**, *31*, 110–123. [CrossRef]

28. Giff, G.; Van Loenen, B.; Crompvoets, J.; Zevenbergen, J. Geoportals in selected European states: A non-technical comparative analysis. In Proceedings of the Conference, Small Island Perspectives on Global Challenges: The Role of Spatial Data in Supporting a Sustainable Future, St. Augustine, Trinidad, 25–19 February 2008; pp. 25–29.
29. Lim, E.-P.; Goh, D.H.-L.; Liu, Z.; Ng, W.-K.; Khoo, C.S.-G.; Higgins, S.E. G-Portal: A map-based digital library for distributed geospatial and georeferenced resources. In Proceedings of the 2nd ACM/IEEE-CS Joint Conference on Digital Libraries, Portland, OR, USA, 14–18 July 2002; ACM: New York, NY, USA, 2002; pp. 351–358.
30. Crompvoets, J.; Bregt, A.; de Bree, F.; van Oort, P.; van Loenen, B.; Rajabifard, A.; Williamson, I. Worldwide (Status, Development and) Impact Assessment of Geoportals. FIG Working Week 2005 and GSDI-8. Available online: https://minerva-access.unimelb.edu.au/handle/11343/33848 (accessed on 4 September 2019).
31. Jiang, H.; Van Genderen, J.; Mazzetti, P.; Koo, H.; Chen, M. Current status and future directions of geoportals. *Int. J. Digit. Earth* **2019**. [CrossRef]
32. Veenendaal, B.; Brovelli, M.A.; Li, S. Review of Web Mapping: Eras, Trends and Directions. *ISPRS Int. J. Geo-Inf.* **2017**, *6*, 317. [CrossRef]
33. ESRI. *ESRI Geoportal Technology. An ESRI® White Paper*; ESRI: Redlands, CA, USA, 2009.
34. Katumba, S.; Coetzee, S. Employing Search Engine Optimization (SEO) Techniques for Improving the Discovery of Geospatial Resources on the Web. *ISPRS Int. J. Geo-Inf.* **2017**, *6*, 284. [CrossRef]
35. Iosifescu-Enescu, I.; Matthys, C.; Gkonos, C.; Iosifescu-Enescu, C.M.; Hurni, L. Cloud-Based Architectures for Auto-Scalable Web Geoportals towards the Cloudification of the GeoVITe Swiss Academic Geoportal. *ISPRS Int. J. Geo-Inf.* **2017**, *6*, 192. [CrossRef]
36. Li, Y.; Wei, B.; Wang, X. A Web-Based Visual and Analytical Geographical Information System for Oil and Gas Data. *ISPRS Int. J. Geo-Inf.* **2017**, *6*, 76. [CrossRef]
37. Kumar, L.; Skidmore, A.K.; Knowles, E. Modelling topographic variation in solar radiation in a GIS environment. *Int. J. Geogr. Inf. Sci.* **1997**, *11*, 475–497. [CrossRef]

International Journal of
Geo-Information

HiBuffer: Buffer Analysis of 10-Million-Scale Spatial Data in Real Time

Mengyu Ma, Ye Wu *, Wenze Luo, Luo Chen **, Jun Li and Ning Jing**

College of Electronic Science, National University of Defense Technology, Changsha 410073, China;
mamengyu10@nudt.edu.cn (M.M.); luowenze12@nudt.edu.cn (W.L.); luochen@nudt.edu.cn (L.C.);
junli@nudt.edu.cn (J.L.); ningjing@nudt.edu.cn (N.J.)
* Correspondence: yewugfkd@nudt.edu.cn; Tel.: +86-137-5518-6456

Received: 30 October 2018; Accepted: 27 November 2018; Published: 30 November 2018

Abstract: Buffer analysis, a fundamental function in a geographic information system (GIS), identifies areas by the surrounding geographic features within a given distance. Real-time buffer analysis for large-scale spatial data remains a challenging problem since the computational scales of conventional data-oriented methods expand rapidly with increasing data volume. In this paper, we introduce HiBuffer, a visualization-oriented model for real-time buffer analysis. An efficient buffer generation method is proposed which introduces spatial indexes and a corresponding query strategy. Buffer results are organized into a tile-pyramid structure to enable stepless zooming. Moreover, a fully optimized hybrid parallel processing architecture is proposed for the real-time buffer analysis of large-scale spatial data. Experiments using real-world datasets show that our approach can reduce computation time by up to several orders of magnitude while preserving superior visualization effects. Additional experiments were conducted to analyze the influence of spatial data density, buffer radius, and request rate on HiBuffer performance, and the results demonstrate the adaptability and stability of HiBuffer. The parallel scalability of HiBuffer was also tested, showing that HiBuffer achieves high performance of parallel acceleration. Experimental results verify that HiBuffer is capable of handling 10-million-scale data.

Keywords: buffer analysis; real-time; visualization-oriented; tile-pyramid; parallel computing

---

1. Introduction

A buffer in a geographic information system (GIS) is defined as the zone around a spatial object, measured by units of time or distance [1]. Buffer analysis is a basic GIS spatial operation for overlay analysis, proximity analysis, spatial data query, and so on. Buffer generation is the core issue in buffer analysis, and several methods for solving the buffer generation problem have been proposed. According to the types of buffers, the methods can be summarized by two categories: raster-based and vector-based buffer generation methods. Raster-based buffer generation methods use the values of the pixels in raster images to indicate buffer zones. The resolution of raster buffers is low while zooming in due to sawtooth distortion (see Figure 1a). Vector-based buffer generation methods use vector polygons to represent buffer results. Figure 2 shows the process of vector buffer generation. Compared with raster buffers, vector buffers use much less space for storage, and zooming in does not sacrifice resolution. However, in practice, circles or circular arcs in vector buffers are simplified to regular polygons or regular-polygon segments to reduce computational complexity. As shown in Figure 1b, distortion occurs in vector buffers while zooming in.

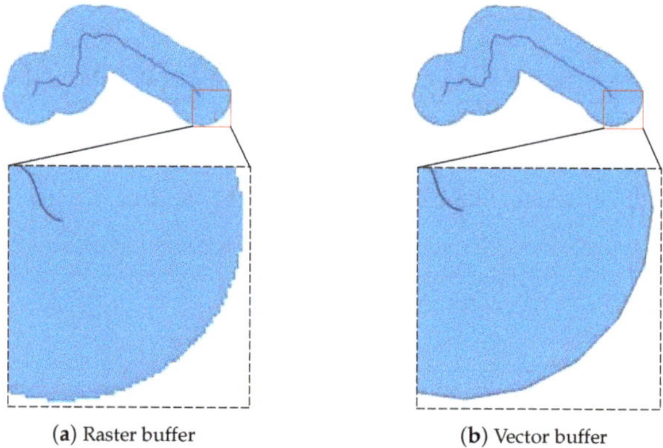

(**a**) Raster buffer   (**b**) Vector buffer

Figure 1. Distortion of raster and vector buffers while zooming in.

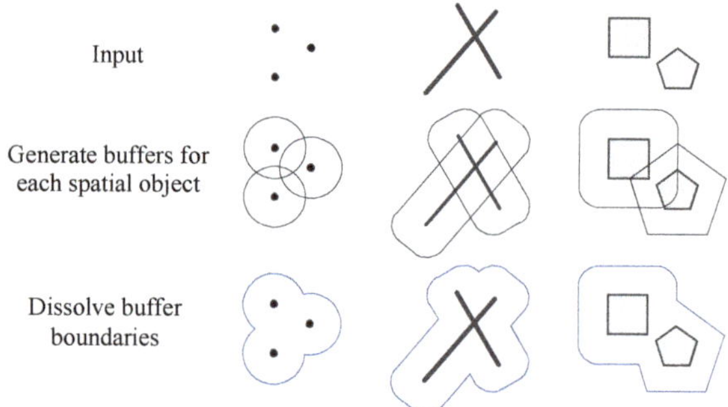

Figure 2. Generating vector buffers.

Buffer generation is computationally demanding. Furthermore, with the rapid development of surveying and mapping technology, spatial data with a larger scale are produced, which will surely increase the computational complexity of buffer generation. Previous studies have proposed many strategies to optimize the problem. Most of the optimization strategies focus on accelerating the construction of buffer zones. For vector-based methods, many approaches have been put forward to address the buffer boundary-dissolving problem [2–4]. Although excellent performance has been achieved, the optimized methods are implemented based on the serial approach. Therefore, the performance is limited when processing large-scale spatial data. Developments in parallel computing technologies provide a prerequisite for high-performance buffer generation, and several parallel strategies have been proposed to solve the bottlenecks [5–8].

In most existing studies, buffers of spatial objects are generated separately first, and then the buffers are merged to get the final results. Such an approach is data-oriented and straightforward. However, the computational scales expand rapidly with the volume of spatial objects; as a result, it is difficult for the traditional data-oriented methods to provide real-time buffer analysis of large-scale spatial data.

In this paper, we present a visualization-oriented parallel buffer analysis model, HiBuffer, to provide an interactive and online buffer analysis of large-scale spatial data. Different from traditional data-oriented methods, the core problem of HiBuffer is to determine whether the pixels for display are in the buffer of a spatial object or not. To the best of our knowledge, the approach is a brand new idea for buffer generation with the following advantages: (1) real-time analysis; (2) unlimited precision; (3) insensitive to data volumes. Many approaches have been proposed to achieve satisfactory performance. One feature of HiBuffer is an efficient buffer generation method, in which spatial indexes and a corresponding query strategy are incorporated. Organized into a tile-pyramid structure, the buffer results of HiBuffer are provided to users with a stepless zooming feature. Moreover, a fully optimized hybrid parallel processing architecture is proposed in HiBuffer to achieve a real-time buffer for large-scale spatial data. Using real-world datasets, we designed and conducted plenty of experiments to evaluate the performance of HiBuffer, including the real-time property, the parallel scalability, and the influence of spatial data density, buffer radius, and request rate.

The remainder of this paper proceeds as follows. Section 2 highlights the related work and literature. In Section 3, the techniques of HiBuffer are described in detail. The experimental results are presented and discussed in Section 4, with an online demonstration of HiBuffer introduced in Section 5. Conclusions are drawn in Section 6.

2. Related Work

The calculation of buffers is an essential operation of GIS. According to the computing architecture, studies on the problem can be classified into two categories: the calculation of buffers using a serial computing model and the construction of buffers in a parallel computing framework.

The first category mainly concentrates on the generation of a buffer zone for each spatial object. Dong [9] introduced a method which utilizes the rotation transform point formula and recursion approach to accelerate buffer generation. Ren [10] proposed a method that reduces the computation by removing the non-characteristic points from the original geometry based on the Douglas–Peucker algorithm. Peng [11] introduced a method based on expansion operations in mathematical morphology, in which the vector data are rasterized and expanded to generate buffers. Wang [12] introduced buffer generation methods based on a vector boundary tracing strategy. Jiechen [13] proposed a method using run-length encoding to encode raster buffers. The method reduces memory occupation while creating the buffers. Zalik [14] presented an algorithm for constructing the geometric outlines of a given set of line segments by using a sweep-line approach. Based on Zalik's method, Sumeet Bhatia [15] presented an algorithm for constructing buffers of vector feature layers and dissolving the buffers based on a sweep-line approach and vector algebra.

With the development of computer hardware, there has been a rapid expansion in processor numbers, thus making parallel computing an increasingly important issue for processing large-scale spatial data. Parallel computing is an effective way to accelerate buffer generation.

Pang [16] proposed a method for buffer analysis based on a service-oriented distributed grid environment. The method includes a master-slave architecture, in which all the entities are assigned to slave computing nodes equally, and then the master node combines the results from all slave nodes to generate the final buffer zones. Huang [5] introduced parallel buffer algorithm which consists of a point-based partition method and a binary-union-tree-based buffer boundary-dissolving method. Fan [6] proposed a parallel buffer algorithm based on area merging to improve the performance of buffer analysis on processing large datasets. Wang [17] proposed a parallel buffer generation method based on the arc partition strategy, which achieves a better load balance performance than the point partition strategy.

In our previous work, we proposed a parallel vector buffer generation method, the Hilbert Curve Partition Buffer Method (HPBM) [8], that takes advantage of the in-memory architecture of Spark [18]. HPBM is based on the Hilbert filling curve to aggregate adjacent objects and thus to optimize the data distribution among all distributed data blocks and to reduce the cost of swapping data among different blocks during parallel processing. Experiments showed that the proposed method outperforms current popular GIS software and existing parallel buffer algorithms. However, HPBM failed to generate buffers for large-scale spatial data in real time.

In summary, all the methods mentioned above are data-oriented and straightforward, with the computational scales expanding rapidly with the volume of spatial objects. It is difficult for the traditional data-oriented methods to provide buffer analysis of large-scale spatial data in real time, though parallel acceleration technologies are adopted.

3. Methodology

In this section, the key technologies of HiBuffer are introduced. Specifically, HiBuffer is used for the buffer analysis of point and linestring objects, which are commonly used to represent points of interest (POI) and roads in maps. In HiBuffer, buffer generation is visualization-oriented; namely, the core problem of HiBuffer is to determine whether the pixels for display are in the buffer of any spatial object or not.

In HiBuffer, we utilize spatial indexes to determine whether a pixel is in the buffers of spatial objects, and accordingly, an efficient buffer generation method named Spatial-Index-Based Buffer Generation (SIBBG) is proposed. The buffers generated by SIBBG are actually raster based. To avoid the sawtooth distortion of these buffers, we propose the Tile-Pyramid-Based Stepless Zooming (TPBSZ) method. To support real-time buffer analysis of large-scale spatial data, parallel computing technologies are used to accelerate computation, and thus we designed the Hybrid-Parallel-Based Process Architecture (HPBPA).

3.1. Spatial-Index-Based Buffer Generation

Spatial indexes are used to organize spatial data. As an efficient tree data structure widely used for indexing spatial data, R-tree was proposed in 1984 [19] and has been thoroughly studied by researchers [20,21]. The spatial queries using R-tree, including the bounding-box query and nearest-neighbor search, have been fully optimized theoretically and practically. In SIBBG, we utilize R-tree to determine whether a pixel is in the buffers of spatial objects.

As shown in Figure 3, the problem in determining whether a pixel P is in the buffers, given a radius R, can be abstracted as determining whether the circle of radius R centered at P intersects with any spatial objects. An intuitive solution to the problem is as follows: (1) calculate the vector buffer $Circ$ of P with radius R; (2) use the INTERSECT operator provided by R-tree to determine whether $Circ$ intersects with any spatial objects. However, there are some disadvantages to this approach. Firstly, as shown in Figure 1b, in fact, $Circ$ is not a circle but a regular polygon; thus, it will lead to a calculation error. Secondly, as R-tree is implemented by grouping nearby objects and representing them with their minimum bounding rectangle in the next higher level of the tree, R-tree works well only for a bounding-box query, rather than queries using other polygon shapes like $Circ$.

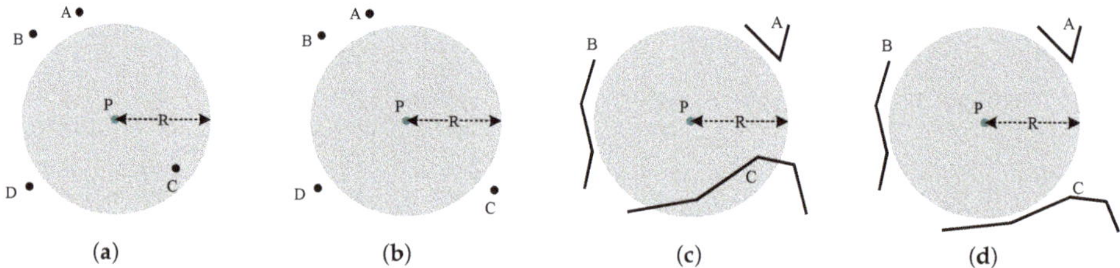

Figure 3. Different situations of pixel P in buffers or not in them, with a given radius R. (**a**) P in buffers of points; (**b**) P not in buffers of points; (**c**) P in buffers of linestrings; (**d**) P not in buffers of linestrings

As the nearest-neighbor search has a much higher computation complexity than the bounding-box query in R-tree, we introduced inner and outer boxes (Figure 4) to optimize the spatial queries in SIBBG. The process of SIBBG is described in Algorithm 1. The inner and outer boxes are used to deal with different situations. In the situation where there are lots of spatial objects within the distance R from P, we query the spatial objects intersecting with the inner box, as a high density of spatial objects in the neighbor is very likely to intersect with the inner box; for the situation where there are few spatial objects in the neighbor of P, we use the outer box to filter out the spatial objects which are far from P. Compared with traditional methods, the performance of SIBBG is less sensitive to the data size.

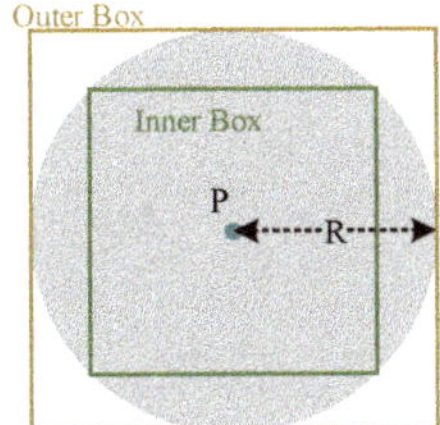

Figure 4. Inner and outer boxes of pixel P with a given radius R.

Algorithm 1 Spatial-Index-Based Buffer Generation.

Input: Pixel P, radius R, and spatial index R-tree.
Output: True or False (whether P is in the buffers of spatial objects with a given radius R).

$r \leftarrow R \times \frac{\sqrt{2}}{2}$
$InnerBox \leftarrow \text{BOX}(P.x - r, P.y - r, P.x + r, P.y + r)$
$Tmp \leftarrow$ satisfying $Rtree.\text{INTERSECT}(InnerBox)$
if Tmp is not *null* **then return** True
else
    $OuterBox \leftarrow \text{BOX}(P.x - R, P.y - R, P.x + R, P.y + R)$
    $Tmp \leftarrow$ satisfy $Rtree.\text{INTERSECT}(OuterBox)$ and $Rtree.\text{NEAREST}(P)$
    if Tmp is not *null* && $\text{DISTANCE}(Tmp, P) \leq R$ **then return** True
return False

3.2. Tile-Pyramid-Based Stepless Zooming

Tile-pyramid is a multi-resolution data structure model widely used for map browsing on the web. The structure of tile-pyramid is shown in Figure 5. At the lowest level of tile-pyramid (level 0), a single tile summarizes the whole map. For each higher level, there are up to 4^z tiles, where z is the zoom level. Each tile has the same size of $n \times n$ pixels and corresponds to the same geographic range [22]. In TPBSZ, a tile-pyramid structure is employed to organize buffer results, provided as a Web Map Tile Service (WMTS). Typically, the tile size is set to 256×256 pixels, which is commonly used in tile-pyramid structures.

The advantages of TPBSZ are mainly reflected by the following two aspects:

1. **(Stepless zooming with unlimited precision)** Tiles of different levels are selected for the screen display according to zoom levels. Zooming in on the buffer results, tiles with higher levels and higher resolutions will be used, and there is no sawtooth distortion. Besides, there is no limit on the max level; in other words, there is no highest resolution limit.

2. **(Stable computational complexity)** The buffer results are accessed by users via a web browser, and only tiles in the screen range need to be generated. As under different zoom levels, the number of tiles in the screen range is limited and stable, the computational complexity remains stable.

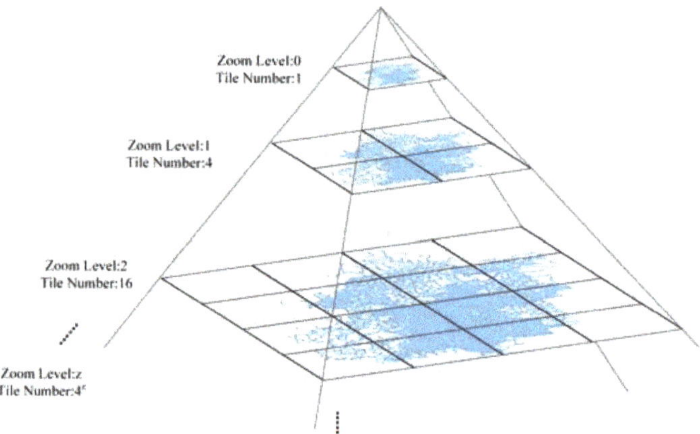

Figure 5. Tile-pyramid structure.

3.3. Hybrid-Parallel-Based Process Architecture

In HiBuffer, hybrid parallel computing technologies are introduced to achieve real-time buffer analysis of large-scale spatial data. The HPBPA of HiBuffer is shown in Figure 6. HPBPA mainly comprises three parts: the **Hybrid-Parallel Buffer Tile Rendering Engine**, **Multi-Thread Buffer Tile Server**, and **In-Memory Messaging Framework**. The **Spatial Indexes Hub** consists of the R-tree indexes of the spatial objects. The R-tree indexes are pre-built quickly and stored in memory-mapped files [23], which will not be totally loaded into the memory. It yields the benefits of less disk I/Os and less memory consumption, even if the indexes are very big. **Task Pool** stores the tasks to be executed. Analysis results are provided to users as a WMTS, which can be browsed through the Internet. The buffer tiles are created when it is requested for the first time, and the tiles are stored in the **Result Pool** to reduce repeated computation if the tiles are requested again.

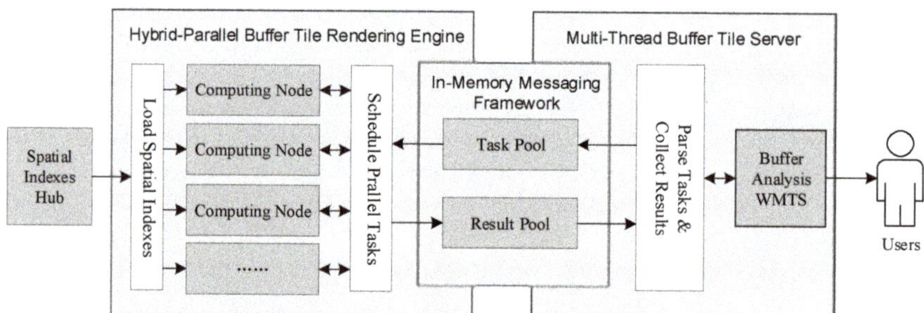

Figure 6. Hybrid-Parallel-Based Process Architecture of HiBuffer.

The **Hybrid-Parallel Buffer Tile Rendering Engine** adopts the hybrid Message Passing Interface (MPI)-Open Multiprocessing (OpenMP) parallel processing model to render buffer tiles. MPI is a multiprocess parallel processing model based on message passing. OpenMP is a multi-thread parallel processing model based on shared memory. In HiBuffer, we treat the rendering of one buffer tile as an independent task, and each task is processed with multiple OpenMP threads in one MPI process. As the task requests are generated by way of streaming, the tasks are dynamically allocated to the MPI processes. An MPI process will be suspended after the assigned task is accomplished, and new tasks will be handled on a first-in-first-served basis. An example of the buffer tile rendering process is shown in Figure 7.

The **Multi-Thread Buffer Tile Server** encapsulates the buffer analysis service as a WMTS. It adds tasks to the **Task Pool** and filters out unnecessary task requests, including (1) tiles for which the distance from Minimum Bounding Rectangle (MBR) of the spatial objects are larger than the given buffer radius; (2) tiles generated in previous tasks which are still in the **Result Pool**. Buffer tiles are returned to users from the **Result Pool** once finished. In order to improve concurrency, multi-thread technology is adopted in the tile server.

The **In-Memory Messaging Framework** is a messaging framework based on Redis, which is an In-Memory Key-Value database. In this messaging framework, tasks and results are transferred rapidly in memory without disk I/Os. The tasks are stored in a FIFO queue in Redis. Tasks are pushed to the queue and popped to suspended MPI processes. To avoid errors in parallel processing, the push and pop operations are performed in blocking mode. After a task is finished, the buffer tile is written to Redis, and a task completion message will be sent to the tile server using subscribe/publish functions in Redis. Tiles are set with expired times, and expired tiles are cleaned up once the max memory limit is exceeded.

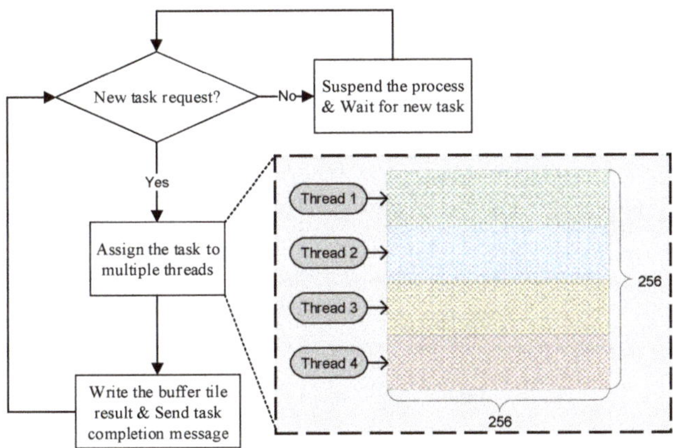

Figure 7. Buffer tile rendering in a Message Passing Interface (MPI) process with four Open Multiprocessing (OpenMP) threads.

4. Experimental Evaluation

In this section, we report several experiments we conducted to evaluate the performance of HiBuffer. First, the ability of HiBuffer to support real-time buffer analysis of large-scale data was tested. Then, we carried out experiments to analyze the influence of different factors on HiBuffer, including spatial data density, buffer radius, and request rate. Finally, the parallel scalability of HiBuffer was tested by running it with varying numbers of MPI processes and OpenMP threads.

4.1. Experimental Setup and Datasets

All the experiments were conducted in an SMP server, shown in Table 1. The code of HiBuffer was implemented using C++ language. The experiments are based on MPICH 3.4, Boost C++ 1.64, and Geospatial Data Abstraction Library 2.1.2. Table 2 shows the datasets used in the experiments. L_1, L_2, L_3, L_4, and P_1 are from OpenStreetMap, which is a digital map database built through crowdsourced volunteered geographic information. The larger datasets, L_5 and P_2, were provided by map service providers. It is worth noting that all the datasets were collected from the real world and have a geographically unbalanced distribution property. Such datasets create challenges with respect to efficient processing.

In the experiments, a task refers to the request of a buffer tile which is in the MBR of spatial objects and also is not in the **Result Pool**. The rendering time of a tile refers to only the time when the tile is rendered in the **Hybrid-Parallel Buffer Tile Rendering Engine**, not including the waiting time while the task is in the **Task Pool**; the rendering time of N tiles ($N > 1$) refers to the whole time cost of rendering N tiles in HiBuffer, which means from the time N tasks are generated in the **Task Pool** until all the tasks have been finished. The experimental settings are listed in Table 3. The benchmark buffer radius was set to 200 m, which has practical meaning for the datasets with a wide spatial range. The benchmark request rate was set to infinity, which means that all the task requests are dispatched simultaneously. As there are 32 cores/64 threads in the server processors, the benchmark parallel processing was set to run with 32 MPI processes and 2 OpenMP threads in each process.

Table 1. Experimental environment.

Item	Description
CPU	32 cores*2, Intel(R)Xeon(R)E5-4620@2.60 GHz
Memory	256 GB
Operating System	Centos 7.1

Table 2. Datasets used in the experiment.

Dataset	Abbreviation	Records	Size
OSM Beijing roads	L_1	40,927	203,413 segments
OSM Taiwan roads	L_2	208,067	2,336,250 segments
OSM Switzerland roads	L_3	597,829	6,991,831 segments
OSM Spain roads	L_4	3,132,496	42,497,196 segments
China roads	L_5	21,898,508	163,171,928 segments
OSM Spain points	P_1	355,105	355,105 points
China points	P_2	20,258,450	20,258,450 points

Table 3. Experimental settings.

Experiments	Buffer Radius (m)	Request Rate (tiles/s)	MPI Processes	OpenMP Threads
1	200	INF	32	2
2	200	INF	32	2
3	200/400/600/800/1000/100,000	INF	32	2
4	200	50/100/200/400/800/INF	32	2
5	200	INF	1/2/4/8/16/32/64	1/2/4

4.2. Experiment 1: Support Real-Time Buffer Analysis of Large-Scale Data with HiBuffer

In order to highlight the superiority of HiBuffer, Table 4 shows a comparison of HiBuffer with HPBM, three optimized parallel methods, and the popular GIS software programs PostGIS, QGIS, and ArcGIS. HiBuffer was deployed and tested in the same hardware environment. We then carried out experiments on the tile rendering time to further evaluate the real-time characteristic of HiBuffer. For each dataset, we generated 5000 tasks through a test program, which requested different buffer tiles at different zoom levels randomly. We analyzed the tile rendering logs, and the experimental results are shown in Figure 8.

Table 4. Performance of traditional data-oriented methods [8] and HiBuffer.

Algorithm	L_1		L_2		L_3	
	Latency(s)	Speedup	Latency(s)	Speedup	Latency(s)	Speedup
HPBM	9.0	/	38.8	/	332.3	/
Method$_1$[a]	15.4	1.7	128.2	3.3	936.9	2.8
Method$_2$[b]	12.3	1.4	75.5	1.9	661.9	2.0
Method$_3$[c]	17.2	1.9	220.8	5.7	2813.4	8.5
PostGIS	34.9	3.9	295.8	7.6	2380.2	7.2
QGIS	129	14.3	2788	71.9	>7200	>7200
ArcGIS	139	15.4	2365	61.0	>7200	>7200
HiBuffer	<1	/	<1	/	<1	/

[a] is based on three optimization methods: split-and-conquer buffer zone generation, vertices-based task decomposition, and tree-like merger [6]. [b] adopts strategies such as vertices-based task decomposition and tree-like merger [5]. [c] uses the equivalent-arc partition strategy [17].

As shown in Table 4, HPBM outperformed the other traditional data-oriented methods. However, HPBM failed to generate buffers for large spatial data in real time. Even for a small dataset, such as L_1, it took 9 s for HPBM to generate the buffer result. As data volumes grew, the computing time using traditional methods increased significantly. Typically, from L_2 to L_3, the data size rose from 2,336,250 segments to 6,991,831 segments (around 3 times), while the computing time using HPBM increased from 38.8 s to 332.3 s (around 8.6 times). The reason for the differing increase rates is that different data distributions led to different buffer zone dissolving complexities. As a result, it is almost impossible to generate buffers for massive spatial data in real time using traditional data-oriented methods. Using HiBuffer, we can get the buffer results for the datasets (L_1, L_2, L_3) in less than 1 s.

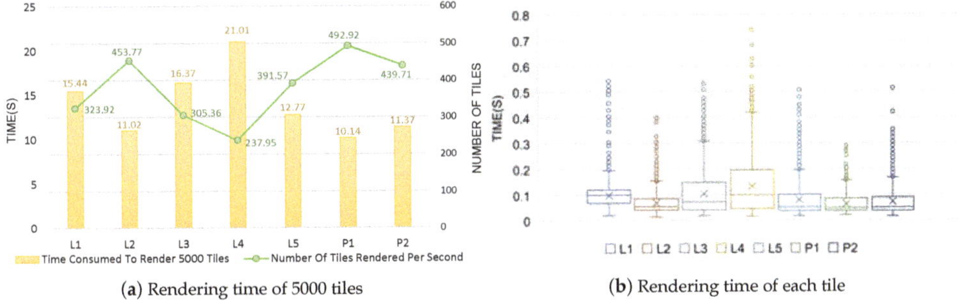

(a) Rendering time of 5000 tiles

(b) Rendering time of each tile

Figure 8. Tile rendering time of HiBuffer on different datasets.

Figure 8a shows the time consumed to render 5000 buffer tiles for each dataset. From L_1 to L_5 or P_1 to P_2, the data size increases sequentially; however, there is no significant uptrend in the rendering time of 5000 buffer tiles. Surprisingly, L_5, the largest dataset with more than 20 million linestring objects, produces better performance than the datasets with much smaller scales (L_1, L_3 and L_4). The experimental results show that HiBuffer is insensitive to data volumes. In Figure 8a, the number of tiles rendered per second is calculated. L_4 produces the poorest performance with 237.95 tiles per second. As the number of tiles in a screen is generally no more than 50, it is possible to perform real-time buffer analysis with HiBuffer for all the datasets. As shown in Figure 8b, the rendering time distributions of each tile on different datasets are visualized with boxplots ('o' represents outliers and '×' represents average rendering time). For all the datasets, L_4 produces the poorest performance though; most of the requested buffer tiles of L_4 are rendered in 0.45 s, with the longest rendering time not exceeding 0.75 s. It is assumed that a browser requests 50 buffer tiles of L_4 at once. Considering that there are 32 MPI processes, the 50 tasks will be processed in two rounds, with 14

$(= 32_{\text{processes}} \times 2 - 50_{\text{tasks}})$ MPI processes suspended in the second round: namely, it will be most likely completed in less than 0.9 s $(= 0.45 \text{ s} \times 2)$. In conclusion, HiBuffer is able to provide interactive and online buffer analysis of large-scale spatial data, which is difficult for traditional data-oriented methods.

4.3. Experiment 2: Impact of Spatial Data Density in HiBuffer

In this experiment, we focused on testing the impact of spatial data density on HiBuffer. Spatial data density refers to the number of spatial elements per unit area. We used the number of segments or points in a tile to signify spatial data density. To express spatial data density exactly, the tile level should be neither too low nor too high. Typically, we chose tiles of level 13, which have proper spatial spans (about 4.9×4.9 km) for all the datasets. For each dataset, we generated 5000 different requests of buffer tiles of level 13 and studied the relationship between tile rendering time and the number of elements in the tiles. The results are shown in Figure 9.

Figure 9a–g illustrate the changes in tile rendering time, along with the increase in element numbers in a tile for each dataset. A point ('o') in the figures represents a sample with a record of rendering time of a tile and number of elements in the tile. The trend lines, generated based on the lowess regression, indicate the trends of the changes. As shown by the trend line of each dataset, the rendering time of a tile has an uptrend, along with the increase in element numbers in a tile, which means the performance decreases with the increase in spatial data density in HiBuffer. This is because higher spatial data density leads to higher computational complexity in the SIBBG process. However, the performance degradation in HiBuffer is not serious (take L_4, for example: when the number of segments in a tile increases to about 28,000, the rendering time of the tile is still less than 0.5 s). The percentage lines illustrate the distributions of 5000 tiles under different spatial data densities, which demonstrate the geographically unbalanced distribution property of the datasets. Figure 9h compares the impact of spatial data density on the different datasets. The average number of elements in the tiles indicates the spatial data density of a dataset, while the average rendering time of each tile indicates the buffer generation performance. As illustrated in the figure, the spatial data density and the performance roughly have the same trend. This shows that datasets with higher spatial data density yield a weaker performance in HiBuffer, which explains the experimental phenomenon that HiBuffer is insensitive to data volumes. L_5 has the largest data volume though, and the lower the spatial data density, the better the performance produced compared with datasets with much smaller scales. Also, the reason that L_1, the small dataset, results in a weaker performance is that the spatial data density is high. The average rendering time of a tile which intersects with no elements was also calculated and is shown in the figure. For different datasets, the average rendering time of a no-element tile always remains at about 0.05 s. This demonstrates that the rendering time of a no-element tile is not related to the spatial data densities or the data volumes of the datasets.

4.4. Experiment 3: Impact of Buffer Radius on HiBuffer

In experiment 3, the impact of the buffer radius on HiBuffer was studied. The buffer radius was set to 200, 400, 600, 800, 1000, and 100,000 m, respectively. For each radius, we generated 5000 buffer tile tasks of different zoom levels for each dataset. We calculated the average rendering time of each tile; the experimental results are shown in Figure 10.

As the buffer radius increases, more pixels in the tiles belong to the buffers of spatial objects, and more computation is required to generate the buffer tiles. This is corroborated by the experimental results: the average rendering time of each tile shows an uptrend with the increase in buffer radius for all the datasets. To estimate the lower-bound performance of HiBuffer, the buffer radius was set to a very large value, 100,000 m, which is 500 times the benchmark buffer radius. The results show that when the buffer radius is 100,000 m, the performance degradation of HiBuffer is not serious. L_4 produces the poorest performance though; the average rendering time of each tile for L_4 is 0.2951 s, which can still support real-time buffer analysis.

(**a**) L_1 (**b**) L_2 (**c**) L_3 (**d**) L_4 (**e**) L_5 (**f**) P_1 (**g**) P_2 (**h**) Impact of spatial data density on different datasets

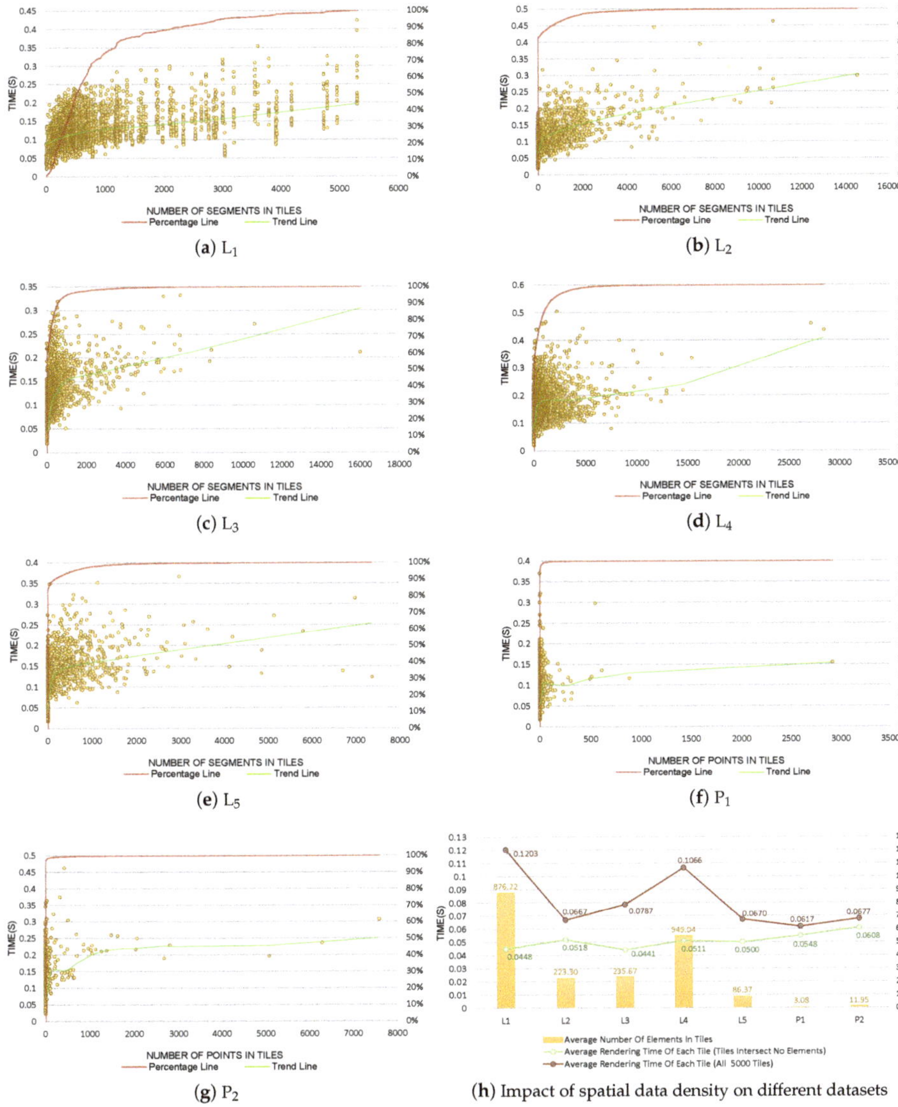

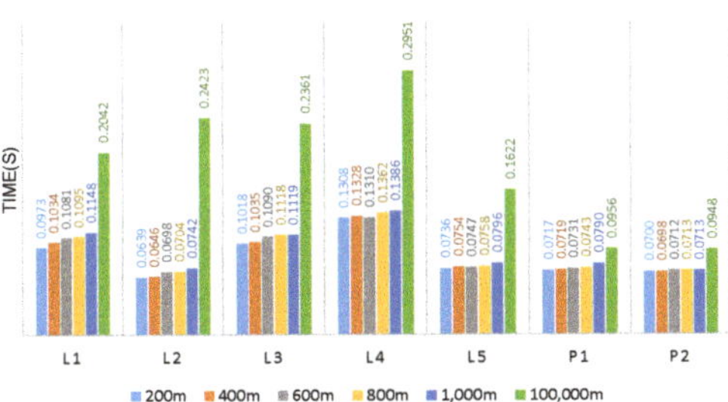

Figure 9. Impact of spatial data density on HiBuffer.

Figure 10. Average rendering time of each tile with different buffer radii in HiBuffer.

4.5. Experiment 4: Impact of Request Rate on HiBuffer

With the exception of experiment 4, the request rate was set to infinity in all other experiments. This means that all the task requests were dispatched simultaneously, and HiBuffer kept running at full load until all tasks were finished. In practical applications, however, the task requests are generated by way of streaming at much lower request rates. In this experiment, the request rate was set to 50, 100, 200, 400, 800, and INF tiles per second, respectively. For each rate, we generated 5000 buffer tile tasks for each dataset.

The rendering time distributions of each tile with different request rates are illustrated in Figure 11. For each dataset, we used the number of tiles rendered per second at the request rate of INF tiles/s (Figure 8a) as the performance limit in HiBuffer. The performance of HiBuffer for all the datasets is affected by the request rate with roughly the same trend. When the request rate is less than the performance limit, the rendering time of a tile increases obviously with the increase in request rates. This results from the intensifying competition for resources between processes in HiBuffer. In contrast, when the request rate exceeds the performance limit, the rendering time of a tile does not change significantly. This is because HiBuffer is running at full load, and the increase in request rates does not cause an obvious effect on the tile rendering performance. However, due to the increase in waiting time while tasks are in the **Task Pool**, the performance of HiBuffer decreases rapidly with the increase in request rate when the request rate exceeds the performance limit. In all the experiments, other than experiment 4, the request rate was set to infinity, which indicates that compared with the experimental results, a higher performance can be achieved in practical applications as a result of the lower request rates.

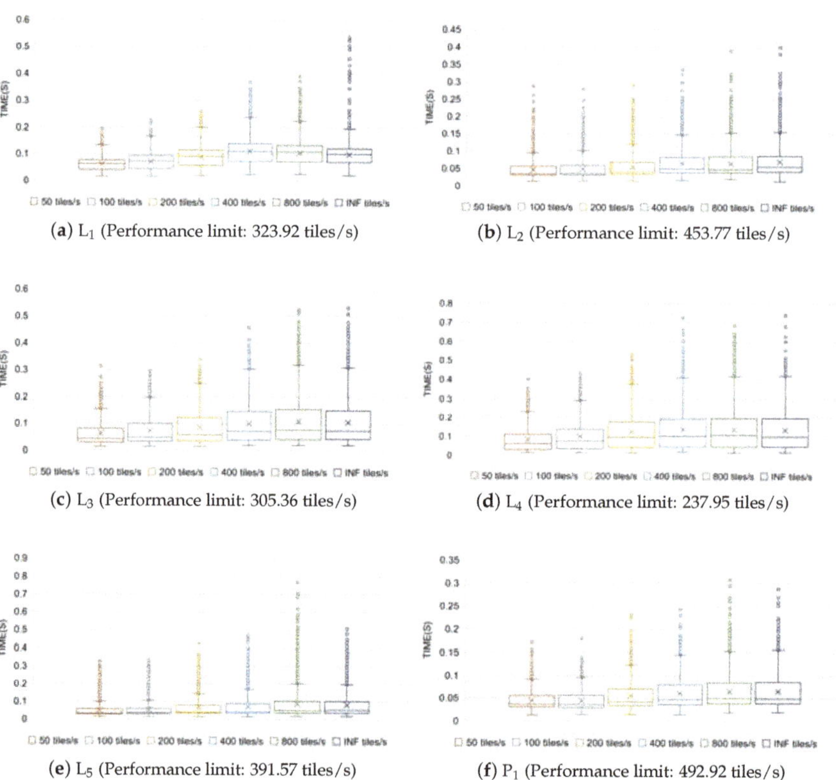

(**a**) L_1 (Performance limit: 323.92 tiles/s)

(**b**) L_2 (Performance limit: 453.77 tiles/s)

(**c**) L_3 (Performance limit: 305.36 tiles/s)

(**d**) L_4 (Performance limit: 237.95 tiles/s)

(**e**) L_5 (Performance limit: 391.57 tiles/s)

(**f**) P_1 (Performance limit: 492.92 tiles/s)

Figure 11. *Cont.*

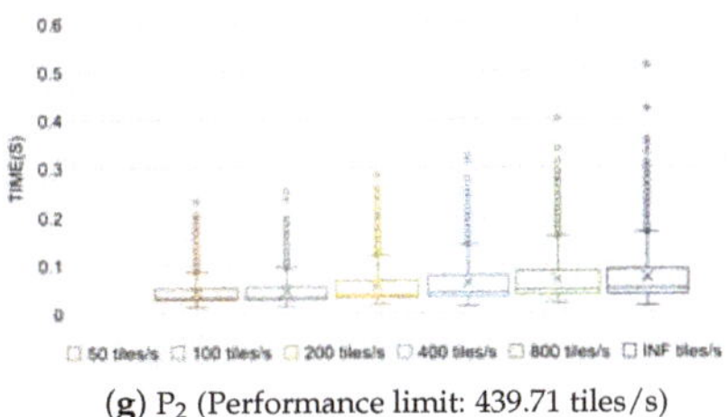

(**g**) P_2 (Performance limit: 439.71 tiles/s)

Figure 11. Rendering time of each tile with different request rates in HiBuffer.

4.6. Experiment 5: Parallel Scalability of HiBuffer

To evaluate the parallel scalability, HiBuffer was respectively tested to run on 1, 2, 4, 8, 16, 32, and 64 MPI processes with 1, 2, and 4 OpenMP threads in each process. For each pair of MPI processes and OpenMP threads, we generated 1000 buffer tile requests of different zoom levels for each dataset. The experimental results are plotted in Figure 12.

First, we analyzed the rendering time of 1000 tiles. The results show that with the increase in process numbers, HiBuffer achieves a high performance of parallel acceleration, which is approximate to linearity when the process number is below 16. However, the performance of parallel acceleration decreases as the process number is increased past 16, especially while running with 4 OpenMP threads in each process. This is because the increase in process numbers aggravates resource competition, which is even worse while running with multi-thread. For example, in Figure 12d, the rendering time of 1000 tiles with 4 threads increases even as the process numbers increase from 16 to 32. Then, we compared and analyzed the average rendering time of each tile line with different OpenMP threads. As shown in the figures, multi-thread parallel processing can effectively reduce the rendering time of a tile when resource competition is not intense. Surprisingly, multi-thread parallel processing even causes performance degradation when the process number is over 16 as a result of the resource competition.

Based on the experimental results and the analysis, we can draw some conclusions on the deployments of HiBuffer in the given hardware environment: (1) in the condition of high load, 64 processes × 1 thread is suggested, because it takes the least time to generate 1000 tiles for all the datasets; (2) in the condition of low load, 16 processes × 4 threads is suggested, as this setting has a high rendering speed of each tile and can reduce the response time of buffer tile requests while the numbers of requests are low.

5. Online Demo of HiBuffer

An online demonstration of HiBuffer is provided on the Web (http://www.higis.org.cn:8080/hibuffer). The OSM Spain roads and points (L_4 and P_1) were used as test data. The demonstration was deployed on a server with four cores/eight threads (see Table 5) and set to run with four MPI processes and eight OpenMP threads. Compared with the experimental environment (see Table 1), there are much fewer CPU cores and much less memory space in the hardware environment of the demonstration, which will surely result in a weaker buffer analysis performance. Even so, as illustrated in the demonstration, it is still possible to provide an interactive buffer analysis even for L_4, the dataset which produces the poorest performance in HiBuffer. Figure 13 shows the analysis results of the online demo.

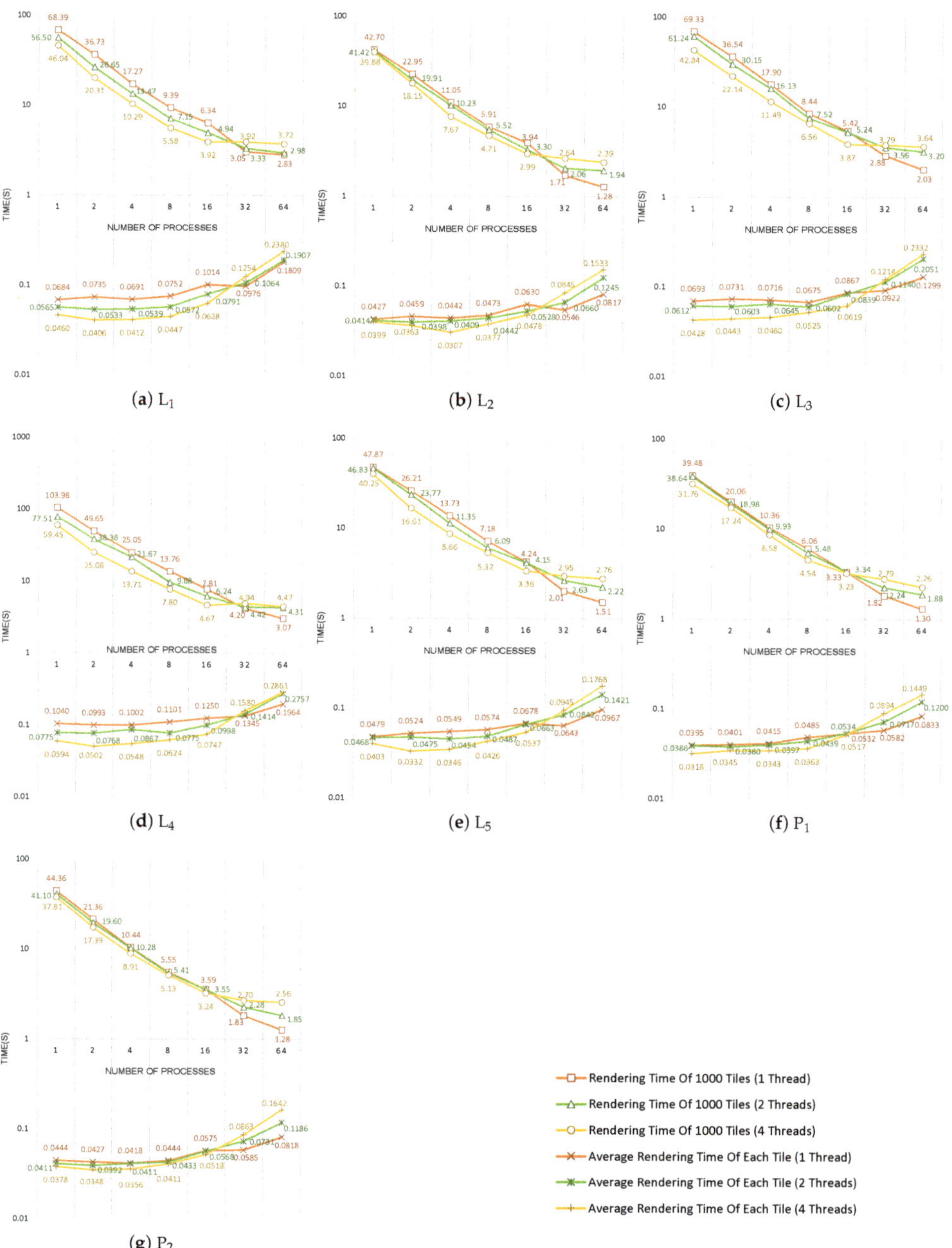

Figure 12. Parallel performance of HiBuffer with different numbers of MPI processes and OpenMP threads.

Table 5. Environment of the online demo.

Item	Description
CPU	4 cores*2, Intel(R)Xeon(R)E5-2680@2.50 GHz
Memory	32 GB
Operating System	Centos 7

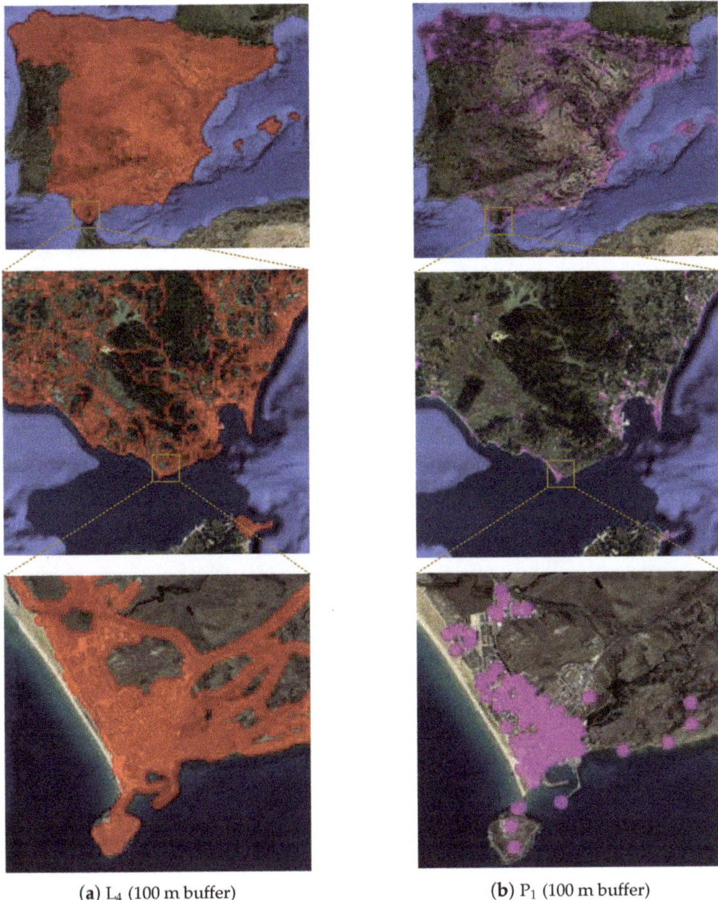

(a) L_4 (100 m buffer) (b) P_1 (100 m buffer)

Figure 13. Analysis results of the online demo.

6. Conclusions and Future Work

This paper presents a visualization-oriented parallel model, HiBuffer, for real-time buffer analysis of large-scale spatial data. In HiBuffer, we propose an efficient buffer generation method named SIBBG. R-tree indexes are utilized in SIBBG to improve buffer generation speed. The performance of SIBBG is not sensitive to the data size. We used the tile-pyramid to organize buffer results and thus propose the TPBSZ. TPBSZ produces satisfactory visualization effects of stepless zooming with unlimited precision and stable computational complexity. Parallel computing technologies are used to accelerate analysis, and we propose the HPBPA. In HPBPA, we designed the fully optimized Hybrid-Parallel Buffer Tile Rendering Engine, Multi-Thread Buffer Tile Server, and In-Memory Messaging Framework.

Our experimental results show that HiBuffer has significant advantages compared with data-oriented methods. Experiment 1 demonstrates the ability of HiBuffer to provide interactive and online buffer analysis of large-scale spatial data. Experiment 2 shows that the performance decreases with the increase in spatial data density in HiBuffer; however, the performance degradation is not serious. In experiment 3, we analyzed the impact of buffer radius on HiBuffer. It shows that the performance decreases slightly with the increase in buffer radius in HiBuffer, and it is still possible to support real-time buffer analysis when the buffer radius is set to 100,000 m, which is a very large value. Then, experiment 4 analyzes the impact of the request rate on HiBuffer. The result indicates that compared with the experimental results, a higher performance can be achieved in practical applications, as the request rate was set to infinity in all the experiments other than experiment 4. In experiment 5, we tested the parallel scalability of HiBuffer. The results show that HiBuffer achieves a high performance of parallel acceleration when resource competition is not intense. Moreover, an online demonstration of HiBuffer is provided on the Web.

HiBuffer does have limitations, and some directions for future research are worth noting:

1. **(Better computation reduction strategies)** In HiBuffer, all buffer tile levels contain the same information but with different resolutions, which indicates that we can use tiles at lower levels to reduce the computation of rendering high-level tiles.

2. **(Buffer analysis of polygon objects)** The buffer generation of point and linestring objects is well-studied in this paper, and future work will focus on applying HiBuffer to buffer analysis of polygon objects.

3. **(Spatial data at a global scale)** In this study, the performance of HiBuffer was tested on datasets at country scales. In the future, we will apply HiBuffer to buffer analysis of global-scale datasets (e.g., the spatial data of the whole world from OpenStreetMap). To support spatial data at a global scale, more optimizations may be needed. For example, to avoid loading the spatial indexes of the whole dataset while generating a buffer tile of a small region, we can divide the spatial data into several parts and build distributed spatial indexes for each part respectively.

4. **(Support overlay analysis)** Overlay analysis is a time-consuming geometric algorithm which combines the spatial data of two input map layers. The operations of overlay analysis include spatial intersection, difference, and union. In our future work, we will extend HiBuffer to solve overlay analysis problems.

Author Contributions: M.M. and Y.W. designed and implemented the algorithm; M.M. and L.C. performed the experiments and analyzed the data; J.L. and N.J. contributed to the construction of experimental environment; W.L. developed the visible interface of the demonstration system; M.M. wrote the paper and Y.W. helped to improve the language expression.

Funding: This research was funded by National Natural Science Foundation of China grant number 41471321 and number 41871284.

Conflicts of Interest: The authors declare no conflict of interest.

Computer Code Availability: The computer code of HiBuffer is open source, and an online demonstration is provided (https://github.com/MemoryMmy/Hibuffer).

Abbreviations

The following abbreviations are used in this manuscript:

GIS	Geographic Information Systems
HPC	High-Performance Computing
HPBM	Hilbert Curve Partition Buffer Method
POI	Points of Interest

WMTS	Web Map Tile Service
MPI	Message Passing Interface
OpenMP	Open Multiprocessing
MBR	Minimum Bounding Rectangle
FIFO	First In First Out
SMP	Symmetrical Multi-Processing

References

1. Sommer, S.; Wade, T. *A to Z GIS: An Illustrated Dictionary of Geographic Information Systems*; Esri Press: Redlands, CA, USA, 2006; pp. 263–264.
2. Sutherland, I.E.; Hodgman, G.W. Reentrant polygon clipping. *Commun. ACM* **1974**, *17*, 32–42. [CrossRef]
3. Weiler, K.; Atherton, P. Hidden surface removal using polygon area sorting. *ACM SIGGRAPH Comput. Graph.* **1977**, *11*, 214–222. [CrossRef]
4. Liang, Y.D.; Barsky, B.A. An analysis and algorithm for polygon clipping. *Commun. ACM* **1983**, *26*, 868–877. [CrossRef]
5. Huang, X. Parallel Buffer Generation Algorithm for GIS. *J. Geol. Geosci.* **2013**, *2*, 115. [CrossRef]
6. Fan, J.; Ji, M.; Gu, G.; Sun, Y. Optimization approaches to mpi and area merging-based parallel buffer algorithm. *Boletim de Ciências Geodésicas* **2014**, *20*, 237–256. [CrossRef]
7. Fan, J. The Key Techniques of Cloud GIS Based on Hadoop. Ph.D. Thesis, The PLA Information Engineering University, Zhengzhou, China, 2013.
8. Shen, J.; Chen, L.; Wu, Y.; Jing, N. Approach to Accelerating Dissolved Vector Buffer Generation in Distributed In-Memory Cluster Architecture. *ISPRS Int. J. Geo-Inf.* **2018**, *7*, 26. [CrossRef]
9. Dong, P.; Yang, C.; Rui, X.; Zhang, L.; Cheng, Q. An effective buffer generation method in GIS. In Proceedings of the 2003 IEEE International Geoscience and Remote Sensing Symposium (IGARSS'03), Toulouse, France, 21–25 July 2003; Volume 6, pp. 3706–3708.
10. Ren, Y.; Yang, C.; Yu, Z.; Wang, P. A way to speed up buffer generalization by Douglas-Peucker algorithm. In Proceedings of the 2004 IEEE International Geoscience and Remote Sensing Symposium (IGARSS'04), Anchorage, AK, USA, 20–24 September 2004; Volume 5, pp. 2916–2919.
11. Peng, H.; Lian, Y.; Chuan-Yong, Y.; Yan-Lan, W. *Map Algebra*; Wuhan University Press: Wuhan, China, 2006.
12. Wang, J.; Chen, Y.; Li, L. Optimization on boundary tracing algorithm of buffer generation. In Proceedings of the 2008 International Conference on Computer and Electrical Engineering, Phuket, Thailand, 20–22 December 2008; pp. 155–159.
13. Wang, J.; Chen, Y.; Dingtao, S. Optimization of Boundary Tracing Algorithm on Buffer Generation. *Geogr. Geo-Inf. Sci.* **2009**, *25*, 95–98.
14. Žalik, B.; Zadravec, M.; Clapworthy, G.J. Construction of a non-symmetric geometric buffer from a set of line segments. *Comput. Geosci.* **2003**, *29*, 53–63. [CrossRef]
15. Bhatia, S.; Vira, V.; Choksi, D.; Venkatachalam, P. An algorithm for generating geometric buffers for vector feature layers. *Geo-Spat. Inf. Sci.* **2013**, *16*, 130–138. [CrossRef]
16. Pang, L.; Li, G.; Yan, Y.; Ma, Y. Research on parallel buffer analysis with grided based HPC technology. In Proceedings of the 2009 IEEE International Geoscience and Remote Sensing Symposium (IGARSS 2009), Cape Town, South Africa, 12–17 July 2009; Volume 4, p. IV-200.
17. Tuo-Di, W.; Ling-Jun, Z.; Li-Zhe, W.; La-Jiao, C.; Qian-Qian, C. Parallel research and opitmization of buffer algorithm based on equivalent arc partition. *Remote Sens. Inf.* **2016**, 147–152.
18. Apache Spark. 2018. Available online: https://spark.apache.org/ (accessed on 25 October 2018).
19. Guttman, A. *R-Trees: A Dynamic Index Structure for Spatial Searching*; ACM: New York, NY, USA, 1984; Volume 14.
20. Cheung, K.L.; Fu, A.W.C. Enhanced nearest neighbour search on the R-tree. *ACM SIGMOD Rec.* **1998**, *27*, 16–21. [CrossRef]
21. Leutenegger, S.T.; Lopez, M.A.; Edgington, J. STR: A simple and efficient algorithm for R-tree packing. In Proceedings of the 13th International Conference on Data Engineering, Birmingham, UK, 7–11 April 1997; pp. 497–506.

22.   Hwa, L.M.; Duchaineau, M.A.; Joy, K.I. Real-time optimal adaptation for planetary geometry and texture: 4–8 tile hierarchies. *IEEE Trans. Vis. Comput. Graph.* **2005**, *11*, 355–368. [CrossRef] [PubMed]

23.   Fernández, F. Boost Geometry Library. 2018. Available online: https://www.boost.org/doc/libs/1680/libs/geometry/doc/html/index.html (accessed on 25 October 2018).

International Journal of
Geo-Information

Article

Mobility Data Warehouses

Alejandro Vaisman [1,*] and **Esteban Zimányi** [2]

[1] Department of Information Engineering, Instituto Tecnológico de Buenos Aires, Lavardén 315,
C1437FBG Ciudad Autónoma de Buenos Aires, Argentina
[2] Department of Computer & Decision Engineering (CoDE), CP 165/15 Université Libre de Bruxelles,
Avenue F. D. Roosevelt 50, B-1050 Brussels, Belgium; ezimanyi@ulb.ac.be
* Correspondence: avaisman@itba.edu.ar; Tel.: +54-11-3457-4864

Received: 9 January 2019; Accepted: 29 March 2019; Published: 2 April 2019

Abstract: The interest in mobility data analysis has grown dramatically with the wide availability of devices that track the position of moving objects. Mobility analysis can be applied, for example, to analyze traffic flows. To support mobility analysis, trajectory data warehousing techniques can be used. Trajectory data warehouses typically include, as measures, segments of trajectories, linked to spatial and non-spatial contextual dimensions. This paper goes beyond this concept, by including, as measures, the trajectories of moving objects at any point in time. In this way, online analytical processing (OLAP) queries, typically including aggregation, can be combined with moving object queries, to express queries like "List the total number of trucks running at less than 2 km from each other more than 50% of its route in the province of Antwerp" in a concise and elegant way. Existing proposals for trajectory data warehouses do not support queries like this, since they are based on either the segmentation of the trajectories, or a pre-aggregation of measures. The solution presented here is implemented using MobilityDB, a moving object database that extends the PostgresSQL database with temporal data types, allowing seamless integration with relational spatial and non-spatial data. This integration leads to the concept of mobility data warehouses. This paper discusses modeling and querying mobility data warehouses, providing a comprehensive collection of queries implemented using PostgresSQL and PostGIS as database backend, extended with the libraries provided by MobilityDB.

Keywords: mobility; data warehouses; spatiotemporal OLAP; mobility analytics

---

1. Introduction

Moving objects [1] (MOs) are objects (e.g., cars, trucks, pedestrians) whose spatial features change continuously in time. The data produced by MOs (e.g., by using attached devices like GPSs and smartphones) can be analyzed in many ways, for example, to discover mobility patterns [2]. This is called mobility data analysis, a technique that is currently used in many related fields, like traffic management, transportation, car pooling applications, smart cities, etc.

The increasing availability of MO data has triggered the interest in mobility data analysis, which has been growing steadily year after year. Moving object data generally come in the form of long sequences of spatiotemporal coordinates $\langle x, y, t \rangle$. To facilitate analysis, these sequences are split into smaller portions of movement, called trajectories. Trajectories can be represented in two possible ways. A continuous trajectory represents the movement track of an object by means of a sequence of spatiotemporal points occurring within a certain interval, together with interpolation functions that allow computing the (approximate) position of the object at any instant. A discrete trajectory contains only a sequence of spatiotemporal points but no interpolation function can be defined.

Moving object databases are databases that store the positions of MOs at any point in time. Although these databases are appropriate for querying, for example, current movement, by means of queries like "which taxis are within ten minutes from Brussels Central Station", they do not support, per se, complex analytical queries such as "For each month, list the total number of buses running at a speed higher than 60 km per hour for more than 40% of their total route." For this, integration with, for example, data warehousing technologies, is needed, yielding the notion of mobility data warehouses. These are data warehouses that contain MO data that can be analyzed in conjunction with other kinds of data (e.g., spatial data), for instance, a road network, altitude data, and the kind.

To represent MOs, the definition of appropriate data types is needed. The notion of temporal type refers to a collection of data types that capture the evolution over time of base types and spatial types. For instance, temporal integers may be used to represent the evolution in the number of employees in a department. Analogously, a temporal point may represent the evolution in time of the position of a vehicle, reported by a GPS device, which would yield a temporal geometry of type point. Over these kind of data types, MO databases can be implemented. Thus, MobilityDB was developed (the demo of this database can be accessed at http://demo.mobilitydb.eu/, where also the manual with a full description of the data types is available). MobilityDB is a MO database (MOD), that builds on PostGIS (the spatial extension of PostgreSQL), that extends the type system of PostgreSQL and PostGIS with abstract data types (ADTs), in order to represent MO data. These ADTs are based on the notion of temporal types and their associated operations.

1.1. Mobility Data Warehouses

Conventional databases are used to support the day-to-day operations in an organization. The operations over these databases are usually denoted as online transactional processing (OLTP). To support data analysis, the notion of data warehousing was developed. A data warehouse (DW) collects large amounts of data from various data sources and reduces them to a form that can be used to analyze the behavior of an organization. DWs are based on the multidimensional model, which represents data as facts that can be analyzed along a collection of dimensions, composed of levels conforming aggregation hierarchies. The multidimensional model builds on the data cube abstraction, where the axes of the cubes are the dimensions, and the cells contain measure values. Typically, DWs are exploited by means of the online analitycal processing (OLAP) technique, which consists of a collection of operations that manipulate the data cube. The most popular OLAP operations are roll-up, which aggregates measure data along a dimension up to a certain aggregation level; drill-down, which is the inverse of the former; slice, which drops a dimension from the cube; and dice, which selects a sub-cube that satisfies a boolean condition.

Spatial data can represent geographic objects (e.g., mountains, cities), geographic phenomena (e.g., temperature, precipitation), etc. Similarly to conventional databases, spatial databases are typically used for operational applications in many different domains, rather than to support data analysis tasks. Spatial DWs (SDWs), on the other hand, combine spatial databases features and DW technologies, to provide more sophisticated data analysis, visualization, and manipulation capabilities. Thus, a SDW is a DW that is capable of representing, storing, and manipulating spatial data. Usually, to represent the extent of a spatial object, spatial data types are used. A SDW can thus take advantage of the operations associated with spatial data types, to allow queries like "give me the total number of theaters within one kilometer from my current position".

Most DWs (either spatial or not) assume that only facts evolve in time. However, dimension data, like for instance, the category of a product, may also change across time. The most popular approach for tackling this issue is based on the notion of slowly changing dimensions [3]. An alternative approach for this issue is based on the concept of temporal databases. Built-in temporal semantics allow these kinds of databases to manage time-varying information. The combination of spatial and temporal databases and DWs leads to the notion of spatiotemporal DWs. Since MO are in essence spatiotemporal, adding MO data to a DW leads to the notion of mobility DWs. This is, in a nutshell, the problem addressed by this paper: modeling and querying mobility DWs by means of adding temporal types to spatial DWs.

1.2. Contributions and Motivation

The authors of this paper had previously defined the notion of spatiotemporal queries as queries that can be expressed by Klug's relational algebra with aggregation, extended with spatial and moving types [4]. Based on this, the authors defined *spatiotemporal DWs* as DWs that support spatiotemporal queries. The classification in the referred work helps to characterize the different kinds of DW systems in terms of their expressiveness, and it is used in the discussion below.

Many efforts have been published under the concept of trajectory data warehouses [5–7]. These works basically propose storing, in the DW fact tables, aggregate measures (like the number of cars in a given latitude-longitude range), segments of trajectories, or spatiotemporal points, probably together with semantic information. It is therefore clear that these approaches do not qualify as trajectory or, equivalently, spatiotemporal DWs (according to the classification in [4]), since they do not support spatiotemporal queries, given that they do not include moving data types. For example, storing aggregate measures [6], does not allow queries like "total distance covered by all cars in Brussels". Instead, they allow expressing queries like "average number of cars in downtown Brussels on Sunday mornings". These approaches conform, in some sense, traditional spatial DWs. The other proposals (although implementations are not yet reported), aim at querying semantic trajectories, for example "total number of persons going from restaurants to theaters". Even in the case where spatiotemporal points are stored, these represent the raw trajectories rather than moving types, and thus everything must be solved using relational operations. On the other hand, there are systems that implement MO databases, typically SECONDO and Hermes. However, these systems are not aimed at building mobility DWs, therefore, implementing a mobility DW based on them is not a trivial task. Section 6 further elaborates on the discussion above.

The work presented in this paper extends and dramatically improves existing trajectory DW proposals. The implementation of MobilityDB gives the possibility of defining MOs as measures in a DW fact table. Integrating relational warehouse data with MO data allows realizing the notion of spatiotemporal queries defined in [4], as this paper will show through a collection of comprehensive examples. The resulting DW is called Mobility DW. In addition, the problem of conceptual and logical modeling of mobility DWs is also covered here. Throughout this paper, the well-known Northwind DW [8], extended with MO and spatial data, is used as a running example.

Note that although the work presented here is based on a centralized PostgresSQL database management system (DBMS), there are ongoing efforts to develop parallel version of this DBMS, from which MobilityDB (and therefore, the mobility DW studied in this paper) can benefit. Further, in a Big Data scenario, horizontal scalability is crucial. This has encouraged the development of a horizontally-scalable version of Postgres, called Postgres-XL, https://www.postgres-xl.org, which can handle mixed workloads, which include OLTP and OLAP queries, GIS queries, OLTP transactions, and so on. Last, but not least, the approach presented here can be extended to big data hadoop-based environments that are currently available, http://spatialhadoop.cs.umn.edu, http://esri.github.io/gis-tools-for-hadoop.

1.3. Paper Organization

In Section 2, the notion of temporal types is defined and explained. Section 3 describes the implementation of temporal types in MobilityDB. Section 4 studies the modeling of mobility DWs, also providing a background on DW design, for the non-expert readers, while Section 5 presents a comprehensive set of queries to illustrate how a mobility DW can be exploited. Section 6 discusses related work and compares such work against the present proposal, and Section 7 concludes the paper.

2. Background: Temporal Types

To make this paper self-contained, this section presents some basic concepts on temporal types. Many of the concepts and operations can be found in [1,8]. However, to develop MobilityDB, new operations had been defined, and are also explained here. Also, a syntax for these temporal types is introduced here.

Temporal types are functions that map time instants to values from a domain. They are implemented by means of applying a constructor temporal(·). For example, a value of type temporal(integer) is a continuous function $f :$ instant $\rightarrow$ integer. Analogously, a temporal boolean is a function mapping instants to boolean values. In temporal databases terminology, several notions for time can be defined. The most popular ones are valid time and transaction time [9]. The former represents the time period during which a database fact is valid in the modeled reality. The latter is the time period during which a fact stored in the database is considered to be true. These are orthogonal concepts. In the remainder, valid time is considered.

Temporal types may be undefined during certain time periods. For example, in Figure 1, Alice's salary is undefined between 1 July 2017 and 1 January 2018 (e.g., because she took a six-month leave), and this is denoted '⊥'. As a convention, closed-open intervals are used.

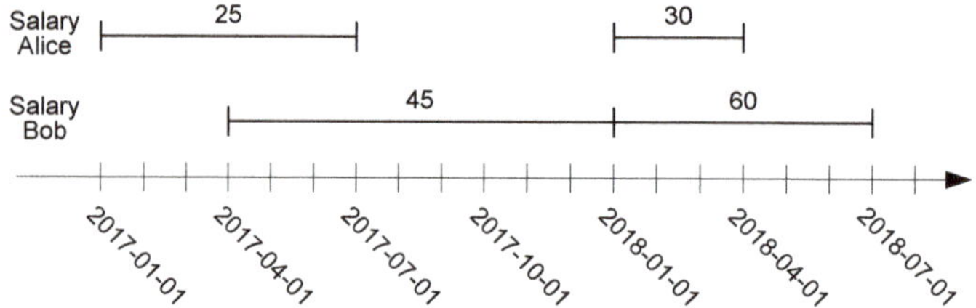

Figure 1. Representation of the evolution of salaries of two employees using the type temporal(integer).

2.1. Classes of Operations on Temporal Types

Temporal types are associated with different kinds of operations, described next.

Projection over domain and range. Operations of this kind are, for example, GetTime and GetValues, which return, respectively, the domain and range of a temporal type.

Interaction with domain and range. Example of operations of this kind are AtTimestamp, AtTimestampSet, AtPeriod, and AtPeriodSet, which restrict the function to a given timestamp (set) or period (set). Operations StartTimestamp and StartValue return, respectively, the first timestamp at which the function is defined and the corresponding value. The operations EndTimestamp and EndValue are analogous. The operations AtValue and AtRange restrict the temporal type to a value or to a range of values in the range of the function. The operations AtMin and AtMax restrict the function to the instants when its value is minimal or maximal, respectively.

Example 1. *The functions GetTime(SalaryAlice) and GetValues(SalaryAlice) return, respectively, {[2017-01-01, 2017-07-01), [2018-01-01, 2018-04-01)} and {25, 30}. Further, AtTimestamp(SalaryAlice, 2017-07-15) returns '⊥', since Alice's salary is undefined at that date. The operation StartTimestamp(SalaryAlice) returns 2017-01-01, while AtValue(SalaryAlice, 25) and AtValue(SalaryAlice, 35) return, respectively, a temporal real with value 20@[2017-01-01, 2017-07-01) and '⊥', because no salary with value 35 exists.*

Temporal aggregation. These kinds of operations are crucial to mobility data analysis. Three basic operations take as argument a temporal integer or real and return a real value namely: Integral, that returns the area under the curve defined by the function, Duration, which returns the duration of the temporal extent over which the function is defined, and Length, which returns the length of the curve defined by the function. From these operations, other derived operations can be defined, such as TWAvg, TWVariance, or TWStDev. The operation TWAvg computes the time-weighted average of a temporal value, taking into account the duration in which the function takes a value. TWVariance and TWStDev compute the variance and the standard deviation of a temporal type. Finally, MinValue and

MaxValue return, respectively, the minimum and maximum value taken by the function. These can be obtained by Min(GetValues($\cdot$)) and Max(GetValues($\cdot$)) where Min and Max are the classic operations over numeric values.

Example 2. *In the example above, TAvg(SalaryAlice) would yield a value of 26.66, given that Alice had a salary of 25 during 181 days and a salary of 30 during 90 days.*

Lifting. This class represents the generalization of the operations on nontemporal types to temporal types [10]. An operation for nontemporal types is lifted to allow any of the arguments to be replaced by a temporal type, and returns a temporal type. For example, the "less than" ($<$) operation has lifted versions (denoted by #$<$) where one or both of its arguments can be temporal types, yielding a temporal boolean. Intuitively, the result is computed at each instant using the nonlifted operation. When two temporal values are defined on different temporal extents, the result of a lifted operation can be defined either over the intersection of both extents or the union of them. In the remainder, for the lifted operations, the first option is assumed.

Example 3. *In Figure 1, the comparison SalaryAlice $<$ SalaryBob results in a temporal boolean with value true@{[2017-04-01, 2017-07-01), [2018-01-01, 2018-04-01)}.*

Lifted aggregation operations are such that the aggregate function is computed at each time instant, and the result is defined over the union of all the extents. Figure 2 shows an exampe of a lifted average for the two salaries in Figure 1.

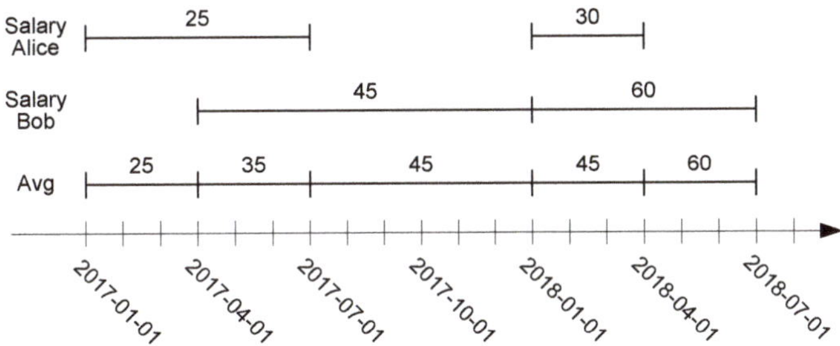

Figure 2. Example of the temporal average operation.

2.2. Temporal Types over Spatial Data

Temporal types can be also defined over spatial data types, not only over basic types like integers or reals. For example, the trajectory of a truck can be represented using a temporal(point) type. Analogously to what was explained above, this type would be a continuous function f : instant $\to$ point. Some of the operations described in Section 2.1 are explained next for the spatial case. The example in Figure 3 shows two temporal points RouteT1 and RouteT2 that represent the delivery routes of two trucks T1 and T2 on a particular day. It can be seen that, for instance, truck T1 took 15 min to go from point (3, 3) to point (0, 0), while truck T2 took 15 min to go from point (4, 0) to point (1, 3). A constant speed between consecutive pairs of points is assumed.

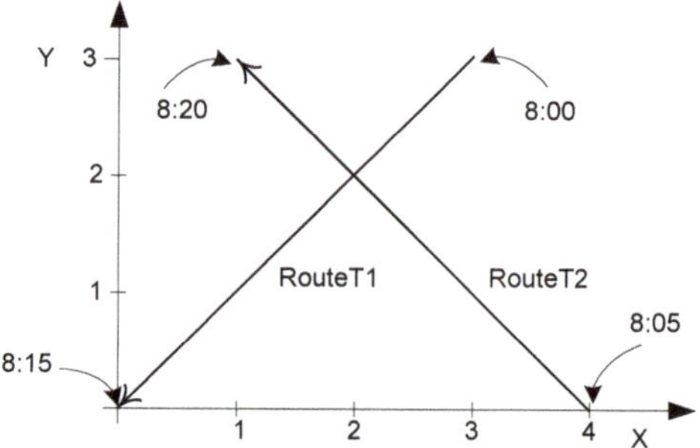

Figure 3. Trajectories of two trucks as a projection of a temporal point into the plane.

An operation called Trajectory, projects a temporal geometry into the plane. In the example, Trajectory(RouteT1) results in the leftmost line in Figure 3, without any temporal information. Note that the projection of a temporal point into the plane may consist in points and lines.

As an example of an operation of the class addressing the interaction with domain and range (as in Section 2.1) for the spatial case, the AtGeometry operation restricts the temporal point to a given geometry. For example, if Polygon denotes a polygon defined as follows Polygon((0 0, 0 2, 2 2, 2 0, 0 0)), then AtGeometry(RouteT1, Polygon) will return the value RouteT1 restricted to the period [8:05, 8:15].

Analogously to the non-spatial case, all operations over nontemporal spatial types are lifted. For example, the Distance function has lifted versions where one or both of its arguments can be temporal points and the result is a temporal real. In the example, Distance(RouteT1, RouteT2) returns a temporal real shown in Figure 4, where, for instance, the value evolves from $\sqrt{8}$@8:05 to 2@8:10 to $\sqrt{8}$@8:15. Note that in this case, the distance function has been approximated by a linear function.

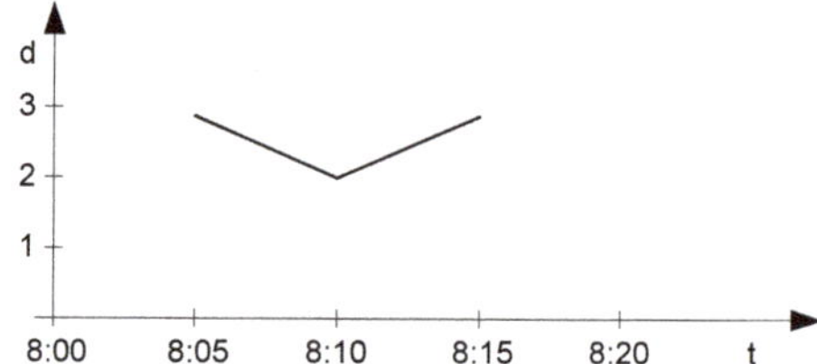

Figure 4. Distance between the trajectories of the two trucks in Figure 3 represented as a temporal(real).

Lifted topological operations return a temporal boolean. Examples of such operators are TIntersects, TDWithin, and TContains. For example, TIntersects(RouteT1, RouteT2) returns a temporal boolean with value false@[8:05, 8:15] since the two trucks were never at the same point at any instant of their route. Similarly, TDWithin(RouteT1, RouteT2, 2) returns a temporal boolean with value {false@[8:05, 8:10), true@[8:10, 8:10], false@(8:10, 8:15]} since the two trucks were only at a distance of two at 8:10. Finally, if Polygon is defined as above, then TContains(Polygon, RouteT1) will return a temporal boolean with value {false@[8:00, 8:05), true@[8:05, 8:15]}.

To conclude, several operations compute the rate of change for points. Operation Speed yields the usual concept of speed of a temporal point at any instant as a temporal real. Operation Direction returns the direction of the movement, that is, the angle between the *x*-axis and the tangent to the trajectory of the moving point. Finally, the Turn operation yields the change of direction at any instant.

3. Temporal Types in MobilityDB

Temporal support has been introduced in the SQL standard, and implemented (in a limited way) in commercial database systems, by means of adding temporality to tables and associating a period with each row. Nevertheless, for mobility applications this approach does not suffice. For such applications, the temporal evolution of the attribute values is needed, along the lines of the data model proposed by Gadia and Nair [11]. This section describes the implementation in MobilityDB, of the temporal types presented in Section 2 as mentioned, a mobility database based on PostgreSQL and PostGIS. For full details, the reader is referred to the manual (http://demo.mobilitydb.eu/mobilitydb-manual.pdf).

3.1. Types in MobilityDB

In order to manipulate temporal types, MobilityDB uses the timestamptz (a shorthand for timestamp with time zone) type provided by PostgreSQL, and three new types: period, timestampset, and periodset. The period type is a specialized version of the tstzrange (short for timestamp with time zone range) type provided by PostgreSQL. Its functionality is similar to the one of type tstzrange, but with a more efficient implementation. A value of the period type has two bounds, the lower bound and the upper bound, which are timestamptz values. The bounds can be inclusive (represented by "[" and "]"), or exclusive (represented by "(" and ")"). A period value with equal and inclusive bounds corresponds to a timestamptz value. An example of a period value is as follows

```
SELECT period '[2017-01-01 08:00:00, 2017-01-03 09:30:00)';
```

The timestampset type represents a set of distinct timestamptz values. A timestampset value must contain at least one element, in which case it corresponds to a timestamptz value. The elements composing a timestampset value must be ordered. An example of a timestampset value is as follows

```
SELECT timestampset '{2017-01-01 08:00:00, 2017-01-03 09:30:00}';
```

Finally, the periodset type represents a set of disjoint period values. A periodset value must contain at least one element, in which case it corresponds to a period value. The elements composing a periodset value must be ordered. An example of a periodset value is as follows

```
SELECT periodset '{[2017-01-01 08:00:00, 2017-01-01 08:10:00],
      [2017-01-01 08:20:00, '2017-01-01 08:40:00]}';
```

Currently, MobilityDB provides six built-in temporal types, tbool, tint, tfloat, ttext, tgeompoint, and tgeogpoint, which are, respectively, based on the bool, int, float, and text types provided by PostgreSQL, as well as the geometry and geography types provided by PostGIS (the last two types restricted to 2D and 3D points). Temporal types may be discrete or continuous. Discrete temporal types (which are based on the boolean, int, or text types) evolve in a stepwise manner, while continuous temporal types (which are based on the float, geometry, or geography types) evolve in a continuous manner. The duration of a temporal value indicates the temporal extent at which the evolution of values is recorded.

Temporal values come in four durations, namely, instant, instant set, sequence, and sequence set. A temporal instant value represents the value at a time instant, such as

```
SELECT tfloat '17.1@2018-01-01 08:00:00';
```

A temporal instant set value represents the evolution of the value at a set of time instants, where the values between these instants are unknown. For example:

```
SELECT tfloat '{17.1@2018-01-01 08:00:00, 17.5@2018-01-01 08:05:00, 18.1@2018-01-01 08:10:00}';
```

A temporal sequence value represents the evolution of the value during a sequence of time instants, where the values between these instants are interpolated using either a stepwise or a linear function. An example is as follows:

```
SELECT tint '(10@2018-01-01 08:00:00, 20@2018-01-01 08:05:00, 15@2018-01-01 08:10:00]';
```

As can be seen, a value of a sequence type has a lower and an upper bound that can be inclusive (represented by '[' and ']') or exclusive (represented by '(' and ')'). The value of a temporal sequence is interpreted by assuming that the period of time defined by every pair of consecutive values v1@t1 and v2@t2 is lower inclusive and upper exclusive, unless they are the first or the last instants of the sequence and in that case the bounds of the whole sequence apply. Furthermore, the value taken by the temporal sequence between two consecutive instants depends on whether the subtype is discrete or continuous. For example, the temporal sequence above represents that the value is 10 during (2018-01-01 08:00:00, 2018-01-01 08:05:00), 20 during [2018-01-01 08:05:00, 2018-01-01 08:10:00), and 15 at the end instant 2018-01-01 08:10:00. On the other hand, the following temporal sequence

```
SELECT tfloat '(10.1@2018-01-01 08:00:00, 20.2@2018-01-01 08:05:00, 15.2@2018-01-01 08:10:00]';
```

represents that the value evolves linearly from 10 to 20 during (2018-01-01 08:00:00, 2018-01-01 08:05:00) and evolves from 20 to 15 during [2018-01-01 08:05:00, 2018-01-01 08:10:00].

Finally, a temporal sequence set value represents the evolution of the value at a set of sequences, where the values between these sequences are unknown, for example:

```
SELECT tfloat '{[17.2@2018-01-01 08:00:00, 17.5@2018-01-01 08:05:00],
        [18.2@2018-01-01 08:10:00, 18.5@2018-01-01 08:15:00]}';
```

3.2. Using Temporal Types

The operations for temporal types defined in Section 2 can be expressed in MobilityDB as explained next. For this, the following table definition is used:

```
CREATE TABLE Employee (
        SSN CHAR(9) PRIMARY KEY,
        FirstName VARCHAR(30),
        LastName VARCHAR(30),
        BirthDate DATE,
        SalaryHist TINT );
```

Tuples can be inserted in this table as follows:

```
INSERT INTO Employee VALUES
( '123456789', 'Alice', 'Cooper', '1980-01-01',
        TINT '{[25@2017-01-01, 25@2017-07-01), [30@2018-01-01, 30@2018-04-01)}'),
( '345345345', 'Bob', 'Brown', '1985-07-25',
        TINT '[45@2017-04-01, 60@2018-01-01, 60@2018-04-01)');
```

The values for the SalaryHist attribute above corresponds to those in Figure 1.

Given the above table with the two tuples inserted, the query

```
SELECT GetTime(E.SalaryHist), GetValues(E.SalaryHist)
FROM    Employee E
```

returns the following values

```
{[2017-01-01, 2017-07-01), [2018-01-01, 2018-04-01)} {25,30}
{[2017-04-01, 2018-07-01)}                            {45,60}
```

The first column of the result above is of type periodset, while the second column is of type integer[] (array of integers) provided by PostgreSQL. Similarly, the query

```
SELECT ValueAtTimestamp(E.SalaryHist, '2017-04-15'),
        ValueAtTimestamp(E.SalaryHist, '2017-07-15')
FROM    Employee E
```

returns the following values

```
25   NULL
45   45
```

where the NULL value above represents the fact that the salary of the Alice is undefined on 2017-07-15. The following query

```
SELECT AtPeriod(E.SalaryHist, '[2017-04-01, 2017-11-01)')
FROM   Employee E
```

returns

```
{[25@2017-04-01, 25@2017-07-01)}
{[45@2017-04-01, 45@2017-11-01)}
```

Note that here, the temporal attributes have been restricted to the period given in the query. The next query is an example of aggregation, and asks for the minimum and maximum values, together with the instants or periods when they occurred:

```
SELECT AtMin(E.SalaryHist), AtMax(E.SalaryHist)
FROM   Employee E
```

The result is:

```
{[25@2017-01-01, 25@2017-07-01]}    {[30@2018-01-01, 30@2018-04-01]}
{[45@2017-04-01, 45@2017-10-01)}    {[60@2017-10-01, 60@2018-07-01]}
```

The use of lifted operations in MobilityDB is illustrated next. The following query asks for the time periods where the salary of Alice was lower than the one of Bob.

```
SELECT E1.SalaryHist #< E2.SalaryHist
FROM   Employee E1, Employee E2
WHERE  E1.FirstName = 'Alice' and E2.FirstName = 'Bob'
```

The query returns the temporal boolean value

```
{[t@2017-04-01, t@2017-07-01), [t@2018-01-01, t@2018-04-01)}
```

Note that when Alice's salary is undefined, no comparison is performed. As an example of lifted aggregation, the next query asks for the average salary across time.

```
SELECT AVG(E.SalaryHist)
FROM   Employee E
```

returns

```
{[25@2017-01-01, 25@2017-04-01), [35@2017-04-01, 35@2017-07-01),
    [45@2017-07-01, 45@2018-04-01), [60@2018-04-01, 60@2018-07-01)}
```

Figure 2 shows this result graphically.

To conclude, the next example illustrates how temporal point types can be used. Consider the following table, which stores the routes followed by trucks to deliver products.

```
CREATE TABLE Delivery (
        TruckId CHAR(6) PRIMARY KEY,
        DeliveryDate DATE,
        Route TGEOMPOINT )
```

Next, two tuples are inserted in this table, containing information of two deliveries performed by two trucks T1 and T2 during the same day (as depicted in Figure 3) :

```
INSERT INTO Delivery VALUES
( 'T1', '2017-01-10',
        TGEOMPOINT '[Point(3 3)@2017-01-10 08:00, Point(0 0)@2017-01-10 08:15]' ),
( 'T2', '2017-01-10',
        TGEOMPOINT '[Point(4 0)@2017-01-10 08:05, Point(1 3)@2017-01-10 08:20]' );
```

The routes represent continuous trajectories. Therefore, a constant speed between any two consecutive points is assumed, and linear interpolation for determining the position of the trucks at any instant is used. Examples of lifted spatial operations are given next.

The following query computes the distance between the two trucks at any time instant.

```
SELECT Distance(D1.Route, D2.Route)
FROM   Delivery D1, Delivery D2
WHERE  D1.TruckId = 'T1' AND D2.TruckId = 'T2'
```

This query returns the temporal float depicted in Figure 4 as follows:

```
[2.82842712474619@2017-01-10 08:05:00, 2@2017-01-10 08:10:00,
        2.82842712474619@2017-01-10 08:15:00]
```

Finally, the following query uses the lifted TIntersects topological operation for testing whether or not the two trucks intersect, as follows:

```
SELECT tintersects(D1.Route, D2.Route)
FROM   Delivery D1, Delivery D2
WHERE  D1.TruckId = 'T1' AND D2.TruckId = 'T2'
```

and the result will be

```
[false@2017-01-10 08:05:00, false@2017-01-10 08:15:00]
```

4. Modeling Mobility Data Warehouses

This section studies how DWs can be extended with temporal types in order to support the analysis of mobility data. The well-known Northwind case study is used in order to introduce the main concepts. First, basic concepts about DW modeling are introduced to make this paper self-contained.

4.1. A Short Introduction to Conceptual Modeling of Data Warehouses

The conventional database design process includes the creation of database schemas at the conceptual, logical, and physical levels. Typically, databases are designed at the conceptual level using some variation of the well-known entity-relationship (ER) model, since it has been acknowledged that conceptual models allow better communication between designers and users for the purpose of understanding application requirements.

Opposite to conventional databases, there is no widely accepted conceptual model for multidimensional data. As a consequence, DW design is usually at the logical level, based on star and/or snowflake schemas [3], which are less intuitive for the final user. The present paper adopts the MultiDim model [8] to represent, at the conceptual level all elements required in DW and OLAP applications, that is, dimensions, hierarchies, and facts with their associated measures. The main reason to adopt this model is that it has also been extended to support spatial data. A streamlined description of the main components of the model follows, based on Figure 5, which represents a DW for the well-known Northwind database.

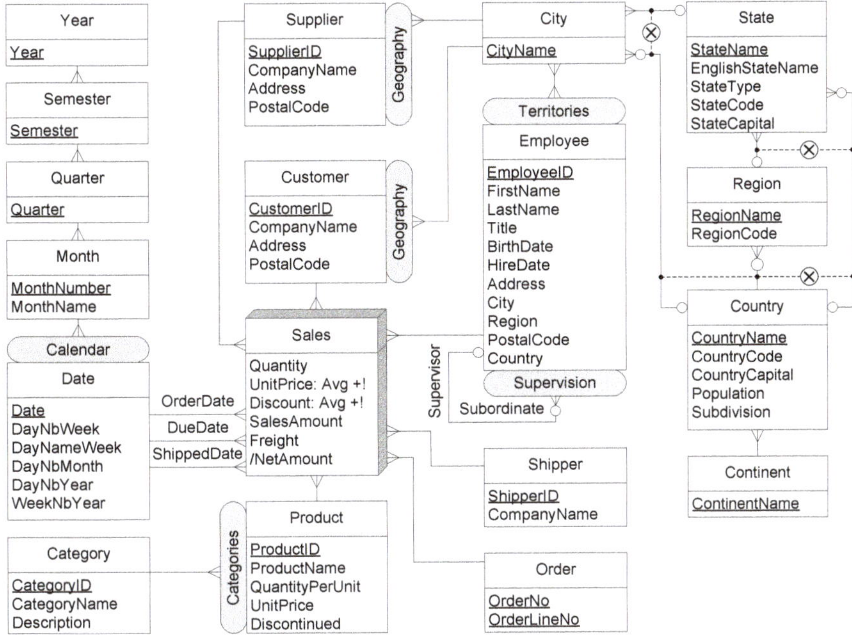

Figure 5. Conceptual schema of the Northwind data warehouse.

A schema is composed of a set of dimensions and a set of facts. A dimension consists of either one level, or one or more hierarchies, and a hierarchy is in turn composed of a set of levels. Instances of a level are called members. For example, Product and Category are levels in Figure 5. As shown in the figure, a level has a set of associated attributes that describe the characteristics of their members, and has one or more identifiers that uniquely identify their members. For example, in Figure 5, CategoryID is an identifier of the Category level. Each attribute of a level has a type, that is, a domain for its values (typically, integer, real, and string). For each pair of levels in a hierarchy, the lower level is called the child and the higher level is called the parent. The relationships composing of hierarchies are called parent-child relationships. The cardinalities of parent-child relationships indicate the minimum and the maximum number of members in one level that can be related to a member in another level. For example, in Figure 5 there is a one-to-many relationship between the child level Product and the parent level Category. Finally, it is sometimes the case that two or more parent-child relationships are exclusive. This is represented using the symbol '⊗'. This is shown in Figure 5, where states can be aggregated either into regions or into countries. Thus, according to their type, states participate in only one of the relationships departing from the State level.

A fact relates several levels. For example, the Sales fact in Figure 5 relates the Employee, Customer, Supplier, Shipper, Order, Product, and Date levels. The same level can participate several times in a fact, playing different roles such that each role is represented by a separate link between the level and the fact. For example, in Figure 5 the Date level participates in the Sales fact with the roles OrderDate, DueDate, and ShippedDate. Instances of a fact are called fact members. The cardinality of the relationship between facts and levels, indicates the minimum and the maximum number of fact members that can be related to level members. For example, in Figure 5 there is a one-to-many relationship between Sales and Product. A fact may contain (usually numeric) attributes commonly called measures, that are analyzed using the context provided by the dimensions. For example, the Sales fact in Figure 5 includes the measures Quantity, UnitPrice, Discount, SalesAmount, Freight, and NetAmount. Levels in a hierarchy are used to analyze factual data at various *granularities*, or levels of detail. In OLAP, this operation is called Roll-up, and aggregates measures along dimension hierarchies, up to a dimension level, using an aggregate function. The aggregation function associated with a measure can be specified next to the measure name, and the SUM aggregation function is assumed by default.

4.2. Conceptual Modeling of Mobility Data Warehouses

The extension of the Northwind DW to support mobility analysis is depicted in Figure 6 and explained below. The idea is aimed at building a mobility DW that keeps track of the deliveries of goods to their customers. In the company there is a fleet of trucks that load the goods in a warehouse (there are several of them), perform a delivery visiting the customers according to a delivery plan, and then return to the warehouse. There is nonspatial data about the customers, the trucks, and the warehouses that store the goods to be delivered to customers. There is also spatial data about the road network (more on this, below). Further, the geographic hierarchies are also represented as spatial data. In addition, there are the trajectories followed by the trucks (derived as a projection of the deliveries). Figure 6 shows the conceptual schema depicting this scenario using the MultiDim model extended to support spatial data and temporal types. This is explained in more detail next.

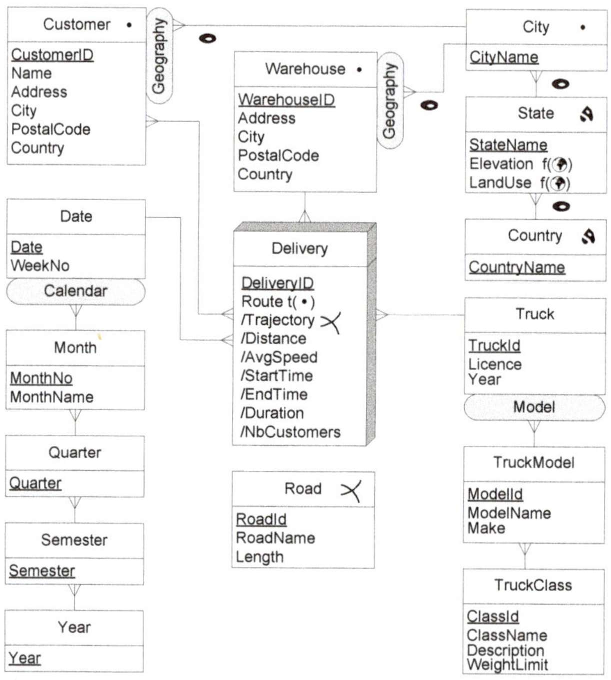

Figure 6. Conceptual schema of the Northwind mobility data warehouse.

As shown in the figure, the fact Delivery is related to four dimensions: Truck, Date, Warehouse, and Customer, where the latter is related to the fact through a many-to-many relationship. The Customer dimension is composed of four levels, with a one-to-many parent-child relationship defined between each pair of levels. The level Truck has attributes TruckId, Licence, and Year. Spatial levels or attributes have an associated geometry (e.g., point, line, or region), which is indicated by a pictogram. In the example, dimensions Customer and Warehouse are spatial and share a Geography hierarchy where a geometry is associated with each level in both dimensions. Also, the State level includes to continuous fields (raster) attributes, namely LandUse and Elevation, which indicate, at each point, its land use (e.g., residential, industrial) and altitude, respectively. Finally, topological constraints are represented using pictograms in parent-child relationships. For example, the topological constraints in dimension Geography indicate that a city is contained in its parent state and similarly for the other parent-child relationships in the hierarchy.

The fact Delivery has an identifier DeliveryID, and eight measures. The first one, Route, keeps the position of the truck at any point in time. It is a *spatiotemporal measure* of type temporal point, as indicated by the symbol 't(•)'. The other measures are derived from Route. Measure Trajectory stores the geometry of the route traversed by the truck, which is a line, without any temporal information. Measures AvgSpeed, Duration, and NbCustomers are numeric, while StartTime and EndTime are timestamps.

In the example of Figure 6, the movement track of the deliveries is represented as a temporal point. In other words, the trajectory is a measure of this fact, allowing deliveries to be aggregated along the different dimensions. Another use of trajectory aggregation identifies "similar" trajectories and merges them in a single class. Thus, queries like "give me the total number of trajectories by class," or "List all the trajectories similar to the one followed by truck T1 on 25 November 2018" are possible.

Finally, note that the road network is not linked to any fact or dimension in the schema. This is a major difference with respect to other approaches that segment trajectories, and link trajectory segments to road segments.

The translation of the conceptual schema in Figure 6 into a snowflake schema is shown in Figure 7. The many-to-many relationship between the Delivery fact and the Customer dimension is represented in table DeliveryCustomer, which contains the keys of the tables Delivery and Customer and the attribute Sequence, which states the order in which the customer were visited. The rest of the translation is straightforward, and it is omitted for the sake of brevity.

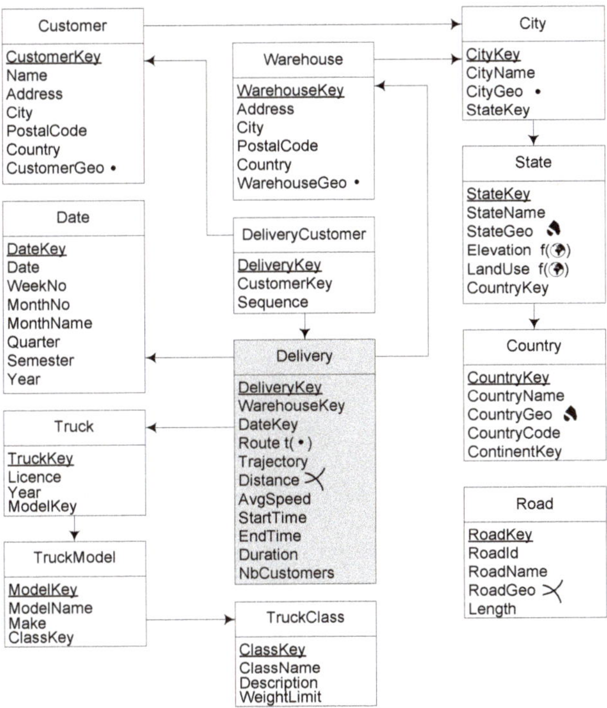

Figure 7. Logical schema of the Northwind mobility data warehouse.

5. Querying the Mobility Data Warehouse in SQL and MobilityDB

In order to express queries to the Northwind mobility DW, the temporal types and their associated operations as defined in Section 3 are used. The mobility DW depicted in Figure 7 is used as an example. Queries were targeted for Belgium, in which the administrative divisions corresponding to states are called provinces. In what follows, the functions that are prefixed by ST_ are PostGIS functions, while the functions not prefixed (e.g., length) are the temporal extensions from MobilityDB. The queries below are aimed at highlighting the advantages of defining MO data as a measure rather than the typical solution of defining static segments of trajectories as measures [5–7]. Therefore, the queries combine typical OLAP operations, like Roll-up, Slice, and Dice, with operations on MOs (in this case, the trucks that perform the deliveries. For each query, the operations involved are explained, to remark the wide variety of queries that the model and implementation allow.

Query 1 (Roll-up + slice + distance). *List the total distance traveled by trucks, per make and month.*
This query involves aggregation along the Date and Truck dimensions, slicing out all other dimensions, and the computation of the distance travelled by the MOs.

```
SELECT    M.Make, DT.Year, DT.MonthNumber, SUM(length(D.Route))
FROM      Delivery D, Truck T, TruckModel M, Date DT
WHERE     D.TruckKey = T.TruckKey AND T.ModelKey = M.ModelKey AND
          D.DateKey = DT.DateKey
GROUP BY M.Make, DT.Year, DT.MonthNumber
ORDER BY M.Make, DT.Year, DT.MonthNumber
```

Also, the query uses the function length to compute the distance traveled by each delivery, and then aggregates the distances per truck make and month in the usual way.

An analysis of this query makes it clear the advantage of the approach presented here, with respect to the approaches that segment the trajectories in order to provide a relational representation of a trajectory. The latter would require, for example, to compute the length of each segment individually (see [8], Chapter 12, for a detailed explanation), and also joining each segment with other dimension tables. Further, proposals that partition the space into a grid, precomputing aggregated measures for each cell in the grid (see Section 6), of course cannot express this query, neither any of the queries in the next examples. Also, even though MO databases like SECONDO and Hermes (also see Section 6) could of course address the MO part of the queries, integrating them into a system to perform OLAP operations is far from being trivial. Therefore, it can be seen that the mobility DW approach based on MobilityDB takes the best of both worlds, bridging the gap between them. The explanation above applies to all queries that are discussed below in this section, thus, it will not be repeated to avoid redundancy.

Query 2 (Roll-up + slice + dice + duration). *List the average duration per day of deliveries starting with a customer located in the province of Namur.*
This query requires aggregation along the Date dimension, a Roll-up operation over the Customer dimension, a Dice operation to select the province of Namur at the State level, and the computation of the duration of the time spanned by the deliveries. Finally, all dimensions but Date are sliced out.

```
SELECT    DT.Date, AVG(duration(D.Route))
FROM      Delivery D, DeliveryCustomer DC, Customer C, State S, Date DT
WHERE     D.DeliveryKey = DC.DeliveryKey AND DC.Sequence = 1 AND
          DC.CustomerKey = C.CustomerKey AND C.StateKey = S.StateKey AND
          S.Name = 'Namur' AND D.DateKey = DT.DateKey
GROUP BY DT.Date
ORDER BY DT.Date
```

The query selects the first customer visited by a delivery using the attribute Sequence, and verifies that she is located in the province of Namur. Then, it uses function duration to compute the duration of the delivery. Finally, the query computes the average of the durations per day.

Query 3 (Roll-up + slice + dice + spatial + duration). *For each month, compute the number of deliveries such that their route intersects the province of Namur for more than 20 min.*

The query performs a roll-up operation along the Date and Customer dimensions, a Dice operation to select the province of Namur at the State level, a spatial operation to compute the intersection, and the computation of the duration of the time spanned by the deliveries within the province. Finally, all dimensions but Date are sliced out.

```
SELECT     T.Year, T.MonthNumber, COUNT(*)
FROM       Delivery D, Date T, State S
WHERE      D.DateKey = T.DateKey AND S.Name = 'Namur' AND D.Route && S.Geom AND
           duration(atGeometry(D.Route, S.Geom)) >= interval '20 min'
GROUP BY T.Year, T.MonthNumber
ORDER BY T.Year, T.MonthNumber
```

This query uses the function atGeometry to restrict the route of the delivery to the geometry of the province of Namur. Then, function duration computes the time spent within the province and verifies that this duration is at least of 20 min. Remark that this duration of 20 min may not be continuous. Finally, the count of the selected deliveries is performed as usual.

The term D.Route && S.Geom is optional and it is included to enhance query performance. It verifies, using an index, that the spatio-temporal bounding box of the delivery projected over the spatial dimension intersects with the bounding box of the geometry. In this way, the atGeometry function is only applied to the deliveries satisfying the bounding box condition.

Query 4 (Roll-up + slice + dice + spatial + duration). *Same as the previous query but with the condition that the route intersects the province of Namur for more than consecutive 20 min.*

The operations involved here are similar to the ones in the previous query. The queries differ in the way in which they compute the duration. Again, this is only possible when measures are actually represented as MOs.

```
SELECT     DT.Year, DT.MonthNumber, COUNT(*)
FROM       Delivery D, Date DT, State S,
           unnest(periods(getTime(atGeometry(D.Route, S.Geom)))) P
WHERE      D.DateKey = DT.DateKey AND S.Name = 'Namur' AND
           D.Route && S.Geom AND duration(P) >= interval '20 min'
GROUP BY DT.Year, DT.MonthNumber
ORDER BY DT.Year, DT.MonthNumber
```

As in the previous query, the function atGeometry restricts the route of the delivery to the province of Namur. This results in a route that may be discontinuous in time, because the route may enter and leave the province. The query uses the function getTime to obtain the set of periods of the restricted delivery, represented as a periodSet value. The function periods converts the latter value into an array of periods, and the PostgreSQL function unnest expands this array to a set of rows, each row containing a period P. Then, it is possible to verify that the duration of one of the periods P has at least 20 min.

Query 5 (Roll-up + slice + distance + speed). *For each month, compute the total number of deliveries that traveled at least 50 km at a speed higher than 100 km/h.*

The operations involved in this query are: a Roll-up along the Date dimension, and the computation of the distance and speed functions on the MO data. Finally, slicing is performed to drop all dimensions but Date.

```
SELECT    DT.Year, DT.MonthNumber, COUNT(*)
FROM      Delivery D, Date DT
WHERE     D.DateKey = DT.DateKey AND length(atPeriodSet(D.Route,
          getTime(atRange(speed(Route) * 3.6, '[100,infinity]')))) / 1000 > 50
GROUP BY DT.Year, DT.MonthNumber
ORDER BY DT.Year, DT.MonthNumber
```

The query uses the function speed to obtain a temporal float representing the speed of the route at each instant, expressed in units per second, depending on the spatial reference system, which in our case is meters per second. This temporal value is then transformed into kilometers per hour by multiplying it by 3.6. The function atRange restricts the speed to the float range [100, infinity]. The getTime function computes the set of periods during which the delivery travels at more than 100 km/h. The function atPeriodSet restricts the delivery to these periods and the length function computes the distance traveled by the restricted deliveries, expressed in meters, which is then converted to kilometers and compared against the value 50.

Query 6 (Temporal aggregation). *For each speed range of 20 km/h, give the total distance traveled by all trucks within that range.*

This is a typical temporal aggregation query, as explained below. That is, operations over MOs are performed over the measure, without involving the dimensions. Only the fact table is used here.

```
WITH Ranges AS (
          SELECT I AS RangeID, floatrange((((I-1)*20), I*20) AS Range
          FROM generate_series(1,10) I )
SELECT    R.Range, SUM(length(atPeriodSet(D.Route,
          getTime(atRange(speed(D.Route) * 3.6, R.Range)))) / 1000)
FROM      Delivery D, Ranges R
WHERE     atRange(speed(D.Route) * 3.6, R.Range) IS NOT NULL
GROUP BY R.RangeID, R.Range
ORDER BY R.RangeID
```

This idea of this query is to allow having an overall view of the speed behaviour of the entire fleet of delivery trucks. The temporary table Ranges stores the ranges $[0, 20), [20, 40), \ldots [180, 200)$. In the main query, the speed of the route, obtained in meters per second, is transformed into kilometers per hour. Then, the function atRange restricts the route to the portions having a given speed range. The time periods comprising the restricted route are obtained with the getTime function. The overall route is restricted then to the obtained time periods with the function atPeriodSet, the distance traveled at the given speed range is obtained with the length function, and all the distances for all trucks at the given time range are obtained with the SUM aggregation function.

Query 7 (Slice + spatial). *Compute the number of deliveries that traversed at least two states.*

The OLAP operation here is a Slice, to drop all dimensions. The rest of the query involves spatial operations to perform intersections between the MOs' trajectories and the State dimension level. Note that, in fact, the intersection is performing a "climbing" along the geographic dimension, skipping the intermediate levels in the hierarchy.

```
SELECT    COUNT(*)
FROM      Delivery D, State S1, State S2
WHERE     S1.StakeyKey <> S2.StateKey AND
          ST_Intersects(trajectory(D.Route), S1.Geom) AND
          ST_Intersects(trajectory(D.Route), S2.Geom)
```

This query projects the route of the delivery to the spatial dimension using the function trajectory, which results in a geometry. The query then tests that the trajectory of the route intersects two states using the PostGIS function ST_Intersects.

Query 8 (Slice + distance + duration). *List the pairs of deliveries that traveled at less than one kilometer from each other during more than half an hour. Again, the OLAP operation here is a Slice, to drop all dimensions. Only the Delivery fact is kept. Then, computations on the temporal types are performed.*

```
SELECT    D1.DeliveryKey, D2.DeliveryKey
FROM      Delivery D1, Delivery D2
WHERE     D1.DeliveryKey <> D2.DeliveryKey AND duration(getTime(
          atValue(tdwithin(D1.Route, D2.Route, 1000), TRUE))) > '20 min'
```

The function tdwithin is the temporal version of the PostGIS function ST_DWithin. It returns a temporal boolean which is true during the periods when the two routes are within the specified distance from each other. The function atValue restricts the temporal boolean to the periods when its value is true, and the function getTime obtains these periods. Finally, the query uses the duration function to obtain the corresponding interval and verifies that it is greater than 20 min.

The next example query involves the LandUse continuous field.

Query 9 (Roll-up + slice + spatial (raster) + distance + duration). *Compute the average duration of the deliveries that started in a residential area and ended in an industrial area on 1 February 2017.*

The OLAP operations are a slice, to drop all dimensions but *Date and Customer, and a roll-up over the Customer (only up to the bottom level) and Date dimensions. Spatial operations over raster data are also performed. Finally, also computations on the temporal types are performed.*

```
SELECT AVG(duration(D.Route))
FROM    Delivery D, Date T, DeliveryCustomer DC1, DeliveryCustomer DC2,
        Customer C1, Customer C2, City Y1, City Y2, State S1, State S2
WHERE D.DateKey = T.DateKey AND D.Date = '2017-02-01' AND
        D.DeliveryKey = DC1.DeliveryKey AND DC1.Sequence = 1 AND
        D.DeliveryKey = DC2.DeliveryKey AND DC2.Sequence = D.NbCustomers AND
        DC1.CustomerKey = C1.CustomerKey AND
        DC2.CustomerKey = C2.CustomerKey AND
        C1.CityKey = Y1.CityKey AND C2.CityKey = Y2.CityKey AND
        Y1.StateKey = S1.StateKey AND Y2.StateKey = S2.StateKey AND
        ST_Intersects(C1.CustomerGeo,AtValue(S1.LandUse, 'Residential')) AND
        ST_Intersects(C2.CustomerGeo,AtValue(S2.LandUse, 'Industrial'))
```

The query starts by selecting the deliveries of the given date. It also selects, using the bridge table DeliveryCustomer, the first and last customers served by the delivery. Then, the query obtains the state of these customers with the subsequent joins. Further, the function AtValue restricts the land use fields of the corresponding states to the values of type residential or industrial. Finally, the function ST_Intersects ensures that the start and end locations of the delivery are included in the restricted fields. Notice that, in PostGIS, the ST_Intersects predicate can compute not only if two geometries intersect, but also if a geometry and a raster intersect.

Note that the query above does not involve temporal data since it neither mentions a temporal geometry such as measure Route, nor a temporal field such as Temperature. The next query involves both temporal attributes, and combines a field and a trajectory.

Query 10 (Slice + spatial (raster) + distance). *List the deliveries that have traveled along more than 50 km of roads at more than 1000 m of altitude.*

A slice is performed to drop all dimensions. Spatial operations over raster data (in this case, representing elevation) are performed. Finally, computations on the temporal types are also performed.

```
SELECT    D.DeliveryNumber
FROM      Delivery D
WHERE     ( SELECT   SUM(ST_Length(ST_Intersection(Trajectory(D.Route),
                     AtRange(S.Elevation, '[1000,infinity)')).geom))
          FROM       State S
          WHERE      ST_Intersects(Trajectory(D.Route), S.StateGeo) ) > 50,000
```

For each delivery, using the function ST_Intersects, the inner query verifies that the route of the delivery and the geometry of the state intersect. Then, the elevation field of the state is restricted to the range of values higher than 1000 m with function AtRange. The function ST_Intersection computes the intersection of the trajectory of the route and the restricted field, and the length of this route is computed with the function ST_Length. The SUM aggregate function is then used to calculate the sum of the lengths of all the obtained routes for all states, and finally the outer query verifies that this sum is greater than 50 km.

Discussion on Performance

Since the paper is oriented to show the features of the mobility data warehousing approach, it is focused on highlighting the expressiveness that MO data adds to the classic and spatial data warehousing approaches. Therefore, a thorough evaluation of query performance is outside the scope of this work. Nevertheless, this section reports a preliminary evaluation, developed as follows. The mobility DW depicted in Figure 7 was implemented over a PostgresSQL relational database. Delivery data (the MO data in attribute Route of the fact table Delivery) had been produced using the data generator of the BerlinMOD benchmark for moving objects http://dna.fernuni-hagen.de/ secondo/BerlinMOD/BerlinMOD.html. However, since the BerlinMOD benchmark targets the city of Berlin, the geographic hierarchy has been dropped, and the references to the State dimension had been replaced by a Borough dimension, both in the schema and in the queries. The evaluation was performed with 520 delivery tuples. The minimum, maximum, and average length of the deliveries are 96, 4632, and 1870, respectively, therefore, this would be approximately equivalent in size to a "normalized" fact table of one million records. There rest of the data are based on the Northwind database. Thus, the table **Customer** contains 1430 tuples, and there are seven warehouses and 100 trucks. Queries 1–8 were run (Queries 9 and 10 were not run, since they include raster data just as an example), changing, as mentioned, the references to the State with references to the Borough. For example, Query 3 now reads:

```
SELECT    T.Year, T.MonthNumber, COUNT(*)
FROM      Delivery D, Date T, Borough B
WHERE     D.DateKey = T.DateKey AND S.Name = 'Mitte' AND D.Route && B.Geom AND
          duration(atGeometry(D.Route, B.Geom)) >= interval '20 min'
GROUP BY T.Year, T.MonthNumber
ORDER BY T.Year, T.MonthNumber
```

Analogously, Query 4 reads, for these experiments:

```
SELECT    DT.Year, DT.MonthNumber, COUNT(*)
FROM      Delivery D, Date DT, Borough B,
          unnest(periods(getTime(atGeometry(D.Route, B.Geom)))) P
WHERE     D.DateKey = DT.DateKey AND B.Name = 'Mitte' AND
          D.Route && B.Geom AND duration(P) >= interval '20 min'
GROUP BY DT.Year, DT.MonthNumber
ORDER BY DT.Year, DT.MonthNumber
```

Table 1 shows the results obtained in these tests. Queries were run several times to eliminate the influence of cold caching (that is, hot chache is assumed). Further, once this influence was eliminated, execution times were the same for each run, therefore just the average value was reported in the table, since the differences between runs were negligible. It can be seen that the queries that took longer were the ones that included spatial operations.

Table 1. Running queries.

Query	Execution Time (s)
Q1	2
Q2	0.35
Q3	180
Q4	180
Q5	12
Q6	60
Q7	60
Q8	9

Of course, these results are not pretended to be conclusive, but suggest that the addition of MOs as fact measures in a warehouse adds query expressiveness. Also, existing proposals segment trajectories (see Section 6) require a "normalization" of the trajectory representation, which is costly due to the number of joins that queries like the ones presented here would require. This is not needed when trajectories are represented as MOs. In addition, note that the DW is implemented using standard object-relational technologies (in this case, PostgresSQL and PostGIS). Therefore, the effect of the inclusion of MOs can also be inferred from an evaluation of the implementation of the temporal types. This is available in the MobilityDB demo site, at URL http://demo.mobilitydb.eu/, where the system runs over the BerlinMOD benchmark data (the queries in such demo do not include warehouse data, they are purely MO queries). It can be observed on the demo site that typical MO queries run very fast over the benchmark, although this is not reported in this paper. Developing a benchmark to perform a comprehensive set of experiments is outside the scope of this paper, and is planned as future work (Section 7).

6. Related Work

Trajectory DWs are strongly related to spatial databases and DWs. The main features of PostGIS, the OGC-based extension to PostgresSQL can be found in [12]. Also, the spatial extension of the MultiDim model presented here was based on the spatial data types of MADS [13], a spatiotemporal conceptual model. The notion of spatial OLAP (SOLAP) was introduced in [14], and it is reviewed in [15]. Other relevant work on SOLAP can be found in [16–18].

In order to support spatio-temporal data, a data model and associated query language for MO data were needed. This is achieved in SECONDO http://dna.fernuni-hagen.de/secondo/ [19], a MO database system developed at the FernUniversität in Hagen, based on the model of Güting et al. [1]. Along similar lines, Hermes is a MOD introduced by Pelekis et al. [20]. The Hermes system is described in [21]. In spite of their ability to handle MO data, neither SECONDO, nor Hermes, are oriented toward addressing the problem of mobility DWs. Among other reasons, integrating both prototypes into existing relational databases is not straightforward. Hermes, for example, extends Oracle through a so-called cartridge with the MO types. However, to build an application on top of the database the application developer must write a source program (for example, in Java), and embed PL/SQL scripts that invoke object constructors and methods from Hermes. SECONDO is as packed system, therefore, integration with existing databases is even more complicated. On the contrary, being coded in the "C" programming language (like PostgresSQL), MobilityDB seamlessly extends the PostGIS library with temporal data types, not requiring any additional software architecture. Thus, building a mobility

DW with MobilityDB turns out to be a natural extension of a spatial relational DW. On the downside, MobilityDB at this time only supports moving points, while SECONDO and Hermes provide (although limited) support to other MOs. However, for traffic analysis, moving points are the most used kind of MOs. The data type system of the temporal types presented in this paper follows the approach of [1]. Also, an SQL extension for spatiotemporal data is proposed in [22].

The work by Orlando et al. [6] introduces the concept of trajectory DWs, aimed at providing the infrastructure needed to deliver advanced reporting capabilities and facilitating the use of mining algorithms on aggregate data. This work is based on the Hermes system. A relevant feature of this proposal is the treatment given to the extraction, transformation and loading (ETL) process, which transforms the raw location data and loads it to the trajectory DW. However, regarding mobility analysis, this proposal does not address MO data. Measures in this trajectory DW are related with the number of MO present in a cell defined by spatiotemporal coordinates. Therefore, queries that can be addressed using this DW are of the form "compute the number of cars between latitudes l_1 and l_2, and longitudes lo_1 and lo_2, on 3 November 2018". However, actual mobility analysis queries like the ones presented in Section 5 cannot be addressed. Also, the authors themselves state in their proposal that MO data analysis remains outside the TDW.

Along similar lines, Wagner et al. [7] proposed the Mob-Warehouse, a conceptual to support mobility analysis using a TDW. The authors propose to enrich trajectory data with semantic information about the domain. The paper is based on the previously defined notion of semantic trajectory [23]. Unlike the model presented in [6], where the measures are pre-aggregated (e.g., number of cars in a cell), in this model the measures are at the finest detail level. No implementation details are given, and MO operations are not addressed. Instead, the proposal is aimed at analyzing semantic trajectories, as the example queries provided suggest.

Based on [7,23], and the work on semantic trajectories, another model for TDWs is proposed in [5], accounting for semantic information. In this case, again, MO data are only partially considered. MO trajectories are partitioned into segments, to which semantic information is associated. No implementation is described, and the authors report than experiments are being carried out, suggesting that the segments in the fact table are not that efficient, which is certainly expected given that many joins are required to reconstruct a trajectory from such segments.

Note that the proposal presented in the present paper does not exclude semantic information. On the contrary, semantic information about MOs can be naturally included in the mobility DW proposed here, and, in fact, this is what dimensions are defined for. Also, all of the features of the proposals above are supported by the model described in this paper (e.g., segmentation, pre-aggregation, semantic information).

An overall perspective of the state of the art in mobility data management can be found in [2,24–26]. A survey on spatiotemporal data warehousing, OLAP, and mining is given in [27]. A discussion about the meaning of spatiotemporal data warehousing is given in [4]. Also, mobility DWs are discussed in [28–30]. Analysis tools for mobility DWs can be found in [31]. Finally, a survey on spatiotemporal aggregation is given in [32], while a survey on trajectory aggregation is provided in [33].

7. Conclusions

This paper studied how DWs can be extended with spatial and mobility data. For this, temporal types were defined to capture the variation of a value across time. The paper discussed how these data types, representing MOs, were implemented as an extension to PostGIS. This extension allows to define a conceptual model for a trajectory DW, which includes MO data as measures that can be analyzed with respect to contextual dimensions. Therefore, the contextual model can actually be implemented with MO data, not only segmented trajectories or pre-aggregated measures. This is a key difference with previous proposals, and would not be possible without a data type system that seamlessly integrates with the (spatial) database supporting the warehouse. A concrete case study that extends the Northwind DW with mobility data was presented, illustrating the use of the mobility

DW with a comprehensive set of queries. Future work includes the development of more kinds of temporal types supporting not only moving points, but also moving lines and moving regions and, as mentioned in Section 5, the development of an appropriate benchmark to allow a fair evaluation of mobility DWs. Another research direction will address the extension of the work presented in this paper, to big data environments, along the lines mentioned in Section 1.

Author Contributions: Both authors contributed equally to the writing of the paper.

Funding: This research received no external funding.

Acknowledgments: Alejandro Vaisman was partially supported by the Argentinian National Scientific Agency, PICT-2014 Project 0787, and PICT-2017 Project 1054.

Conflicts of Interest: The authors declare no conflict of interest.

References

1. Güting, R.; Schneider, M. *Moving Objects Databases*; Morgan Kaufmann: Burlington, MA, USA, 2005.
2. Renso, C.; Spaccapietra, S.; Zimányi, E. (Eds.) *Mobility Data: Modeling, Management, and Understanding*; Cambridge Press: Cambridge, UK, 2018.
3. Kimball, R.; Ross, M. *The Data Warehouse Toolkit: The Complete Guide to Dimensional Modeling*, 2nd ed.; Wiley: Hoboken, NJ, USA, 2002.
4. Vaisman, A.; Zimányi, E. What is spatio-temporal data warehousing? In *Data Warehousing and Knowledge Discovery—Proceedings of the 11th International Conference on Data Warehousing and Knowledge Discovery, DaWaK'09, Linz, Austria, 31 August–2 September 2009*; No. 5691 in Lecture Notes in Computer Science; Springer: Berlin, Germany, 2009; pp. 9–23.
5. Fileto, R.; Raffaetà, A.; Roncato, A.; Sacenti, J.A.P.; May, C.; Klein, D. A Semantic Model for Movement Data Warehouses. In Proceedings of the 17th International Workshop on Data Warehousing and OLAP, Shanghai, China, 3–7 November 2014; pp. 47–56.
6. Orlando, S.; Orsini, R.; Raffaetà, A.; Roncato, A.; Silvestri, C. Spatio-temporal aggregations in trajectory data warehouses. *J. Comput. Sci. Eng.* **2007**, *1*, 211–232.
7. Wagner, R.; de Macedo, J.A.F.; Raffaetà, A.; Renso, C.; Roncato, A.; Trasarti, R. Mob-Warehouse: A Semantic Approach for Mobility Analysis with a Trajectory Data Warehouse. In *Advances in Conceptual Modeling—ER 2018 Workshops, LSAWM, MoBiD, RIGiM, SeCoGIS, WISM, DaSeM, SCME, and PhD Symposium, Hong Kong, China, 11–13 November 2018*; Revised Selected Papers; Springer: Berlin, Germany, 2018; pp. 127–136.
8. Vaisman, A.; Zimányi, E. *Data Warehouse Systems: Design and Implementation*; Springer: Berlin, Germany, 2014.
9. Tansel, A.; Clifford, J.; Gadia, S.; Jajodia, J.; Segev, A.; Snodgrass, T. (Eds.) *Temporal Databases: Theory, Design, and Implementation*; Benjamin-Cummings: San Francisco, CA, USA, 1993.
10. Güting, R.; Böhlen, M.H.; Erwig, M.; Jensen, C.S.; Lorentzos, N.A.; Schneider, M.; Vazirgiannis, M. A foundation for representing and quering moving objects. *ACM Trans. Database Syst.* **2000**, *25*, 1–42.
11. Gadia, S.; Nair, S. Temporal Databases: A Prelude to Parametric Data. In *Temporal Databases: Theory, Design, and Implementation*; Tansel, A., Clifford, J., Gadia, S., Segev, A., Snodgrass, R.T., Eds.; Benjamin-Cummings: San Francisco, CA, USA, 1993; Chapter 2, pp. 28–66.
12. Obe, R.; Hsu, L. *PostGIS in Action*; Manning Publications Co.: Shelter Island, NY, USA, 2014.
13. Parent, C.; Spaccapietra, S.; Zimányi, E. *Conceptual Modeling for Traditional and Spatio-Temporal Applications: The MADS Approach*; Springer: Berlin, Germany, 2006.
14. Rivest, S.; Bédard, Y.; Marchand, P. Toward Better Suppport for Spatial Decision Making: Defining the Characteristics of Spatial On-Line Analytical Processing (SOLAP). *Geomatica* **2001**, *55*, 539–555.
15. Bédard, Y.; Rivest, S.; Proulx, M. Spatial Online Analytical Processing (SOLAP): Concepts, Architectures, and Solutions from a Geomatics Engineering Perspective. In *Data Warehouses and OLAP: Concepts, Architectures and Solutions*; Wrembel, R., Koncilia, C., Eds.; IRM Press: Aarhus, Denmark, 2007; Chapter 13, pp. 298–319.
16. Bimonte, S.; Bertolotto, M.; Gensel, J.; Boussaid, O. Spatial OLAP and Map Generalization: Model and Algebra. *Int. J. Data Wareh. Min.* **2017**, *8*, 24–51.
17. Boulil, K.; Bimonte, S.; Pinet, F. Conceptual model for spatial data cubes: A UML profile and its automatic implementation. *Comput. Stand. Interfaces* **2015**, *38*, 113–132.

18. Viswanathan, G.; Schneider, M. On the Requirements for User-Centric Spatial Data Warehousing and SOLAP. In *Database Systems for Advanced Applications Proceedings of the DASFAA 2011 Workshops, Hong Kong, China, 22–25 April 2011*; Xu, J., Yu, G., Zhou, S., Unland, R., Eds.; Lecture Notes in Computer Science; Springer: Berlin, Germany, 2011; Volume 6637, pp. 144–155.
19. Xu, J.; Güting, R. A generic data model for moving objects. *GeoInformatica* **2018**, *17*, 125–172.
20. Pelekis, N.; Theodoridis, Y.; Vosinakis, S.; Panayiotopoulos, T. Hermes: A Framework for Location-Based Data Management. In *Advances in Database Technology—EDBT 2006—Proceedings of the 10th International Conference on Extending Database Technology, Munich, Germany, 26–31 March 2006*; Ioannidis, Y., Scholl, M., Schmidt, J., Eds.; No. 3896 in Lecture Notes in Computer Science; Springer: Munich, Germany, 2006; pp. 1130–1134.
21. Pelekis, N.; Theodoridis, Y. *Mobility Data Management and Exploration*; Springer: Berlin, Germany, 2014.
22. Viqueira, J.; Lorentzos, N. SQL extension for spatio-temporal data. *VLDB J.* **2007**, *16*, 179–200.
23. Bogorny, V.; Renso, C.; de Aquino, A.R.; de Lucca Siqueira, F.; Alvares, L.O. CONSTAnT—A Conceptual Data Model for Semantic Trajectories of Moving Objects. *Trans. GIS* **2014**, *18*, 66–88.
24. Meng, X.; Ding, Z.; Xu, J. *Moving Objects Management: Models, Techniques and Applications*, 2nd ed.; Springer: Berlin, Germany, 2014.
25. Andrienko, G.; Andrienko, N.; Bak, P.; Keim, D.; Wrobel, S. *Visual Analytics of Movement*; Springer: Berlin, Germany, 2014.
26. Zheng, Y.; Zhou, X. (Eds.) *Computing with Spatial Trajectories*; Springer: Berlin, Germany, 2011.
27. Gómez, L.; Kuijpers, B.; Moelans, B.; Vaisman, A.A. A State-of-the-Art in Spatio-Temporal Data Warehousing, OLAP and Mining. In *Integrations of Data Warehousing, Data Mining and Database Technologies: Innovative Approaches*; Taniar, D., Chen, L., Eds.; IGI Global: Hershey, PA, USA, 2011; Chapter 9, pp. 200–236.
28. Andersen, O.; Krogh, B.B.; Thomsen, C.; Torp, K. An Advanced Data Warehouse for Integrating Large Sets of GPS Data. In Proceedings of the 17th ACM International Workshop on Data Warehousing and OLAP, Shanghai, China, 3–7 November 2014; ACM Press: New York, NY, USA, 2014; pp. 13–22.
29. Marketos, G.; Frentzos, E.; Ntoutsi, I.; Pelekis, N.; Raffaetà, A.; Theodoridis, Y. Building real-world trajectory warehouses. In Proceedings of the 7th ACM International Workshop on Data Engineering for Wireless and Mobile Access, Vancouver, BC, Canada, 13 June 2008; ACM Press: New York, NY, USA, 2008; pp. 8–15.
30. Pelekis, N.; Raffaetà, A.; Damiani, M.-L.; Vangenot, C.; Marketos, C.; Frentzos, E.; Ntoutsi, I.; Theodoridis, Y. Towards trajectory data warehouses. In *Mobility, Data Mining and Privacy: Geographic Knowledge Discovery*; Giannotti, F., Pedreschi, D., Eds.; Springer: Berlin, Germany, 2008; Chapter 9, pp. 189–211.
31. Raffaetà, A.; Leonardi, L.; Marketos, G.; Andrienko, G.; Andrienko, N.V.; Frentzos, E.; Giatrakos, N.; Orlando, S.; Pelekis, N.; Roncato, A.; et al. Visual mobility analysis using T-Warehouse. *Int. J. Data Wareh. Min.* **2011**, *7*, 1–23.
32. Vega López, I.; Snodgrass, R.; Moon, B. Spatiotemporal Aggregate Computation: A survey. *IEEE Trans. Knowl. Data Eng.* **2005**, *17*, 271–286.
33. Andrienko, G.; Andrienko, N. A general framework for using aggregation in visual exploration of movement data. *Cartogr. J.* **2010**, *47*, 22–40.

 International Journal of *Geo-Information*

Article

Parallelizing Multiple Flow Accumulation Algorithm using CUDA and OpenACC

Natalija Stojanovic * and **Dragan Stojanovic**

Faculty of Electronic Engineering, University of Nis, 18000 Nis, Serbia
* Correspondence: natalija.stojanovic@elfak.ni.ac.rs

Received: 29 June 2019; Accepted: 30 August 2019; Published: 3 September 2019

Abstract: Watershed analysis, as a fundamental component of digital terrain analysis, is based on the Digital Elevation Model (DEM), which is a grid (raster) model of the Earth surface and topography. Watershed analysis consists of computationally and data intensive computing algorithms that need to be implemented by leveraging parallel and high-performance computing methods and techniques. In this paper, the Multiple Flow Direction (MFD) algorithm for watershed analysis is implemented and evaluated on multi-core Central Processing Units (CPU) and many-core Graphics Processing Units (GPU), which provides significant improvements in performance and energy usage. The implementation is based on NVIDIA CUDA (Compute Unified Device Architecture) implementation for GPU, as well as on OpenACC (Open ACCelerators), a parallel programming model, and a standard for parallel computing. Both phases of the MFD algorithm (i) iterative DEM preprocessing and (ii) iterative MFD algorithm, are parallelized and run over multi-core CPU and GPU. The evaluation of the proposed solutions is performed with respect to the execution time, energy consumption, and programming effort for algorithm parallelization for different sizes of input data. An experimental evaluation has shown not only the advantage of using OpenACC programming over CUDA programming in implementing the watershed analysis on a GPU in terms of performance, energy consumption, and programming effort, but also significant benefits in implementing it on the multi-core CPU.

Keywords: watershed analysis; parallel processing; multiple flow accumulation; DEM; CUDA; OpenACC; GPU

1. Introduction

Geospatial processing and analysis of large amounts of geospatial data in the Geographic Information System (GIS) represents an application domain that obtains significant benefits from parallel and high-performance computing [1]. Digital Terrain Analysis (DTA) is one of the fundamental features of contemporary GISs used in hydrological modeling, soil erosion, flood simulations, landslide hazard assessment, visibility analysis, telecommunication, and military applications, etc. DTA includes implementation of various algorithms performed by the digital model of the earth surface and terrain, most often Digital Elevation Model (DEM), such as watershed analysis, viewshed analysis, terrain visualization, etc.

Watershed analysis refers to the process of using DEM and iterative operations through raster (gridded) data to delineate watersheds and to detect watershed features such as streams, stream network, drainage divides, basin, outlets, etc. The watershed analysis is used in flood control, erosion detection, and landslide monitoring to provide decision support capabilities.

Watershed analysis consists of several computationally and data intensive tasks (phases) performed over massive DEM data in an iterative manner. As such, it is suitable for accelerating these tasks leveraging parallel and high-performance computing (HPC) methods and techniques.

There are several HPC approaches that can be applied to improve the performance of advanced computationally and data intensive applications, such as the watershed analysis. Such approaches are based on a parallel computing paradigm applied through multi/many-core computer systems, or through distributed computing infrastructure, such as a cluster or cloud infrastructure. While multi-core CPUs are suitable for general-purpose tasks, many-core special purpose processors (Intel Xeon Phi or GPU), comprising of a larger number of lower frequency cores, have been used in general-purpose solutions, performing equally well, or even better in scalable applications. The development of a parallel application and HPC application for such parallel/distributed architectures could be based on various parallel programming frameworks, including OpenACC, OpenCL (Open Computing Language), OpenMP (Open Multi-Processing), and NVIDIA CUDA (Compute Unified Device Architecture). OpenMP is a set of compiler directives, library routines, and environment variables for programming shared-memory parallel computing systems. Furthermore, OpenMP has been extended to support programming of heterogeneous systems that contain CPUs and accelerators. General-purpose computing on a Graphics Processing Unit (GPGPU) represents a new paradigm based on tightly coupled, massively parallel computing units [2]. It represents a method and a technique for performing general purpose computations on a GPU by using an appropriate framework, an API, and a programming model, such as OpenCL, Microsoft's DirectCompute, and NVIDIA CUDA. OpenCL supports portable programming for computer architectures provided by various vendors, while CUDA runs only on NVIDIA Graphics Processing Units (GPU). CUDA combines C/C++ on the host side with C-like kernels that enable general purpose applications to access massively parallel GPUs for non-graphical computing. CUDA C/C++ compiler, libraries, and run-time software enable programmers to develop and accelerate data-intensive applications on GPU. Writing CUDA applications and kernels is a task that is time-consuming and prone to errors requiring detailed knowledge of the target NVIDIA GPU architecture and design of kernel maximally tailored for a particular architecture. In order to provide a parallel programming framework for a large audience and a wide spectrum of multi-core and many-core architectures, an OpenACC standard and parallel programming model has been developed.

OpenACC is a specification of compiler directives and API routines for writing a parallel code in C, C++, and FORTRAN. That code can be compiled into different parallel architectures, such as multi-core CPUs, or many-core parallel accelerators, such as GPUs. In contrast to CUDA where the programmer is required to explicitly decompose the computation into parallel kernels, when using OpenACC, the programmer annotates the existing loops and data structures in the code, so that the OpenACC compiler can target the code to different devices. For NVIDIA GPU devices, the OpenACC compiler generates the kernels, creates the register and shared memory variables, and applies performance optimization.

In this paper, the focus is on accelerating a sequential watershed analysis algorithm using different parallel implementations, a native NVIDIA GPU implementation using CUDA, and OpenACC implementations mapped to both GPU and multi-core CPU. The proposed solutions have been evaluated with respect to the execution time, energy consumption, and programming effort for program parallelization for a different size of input DEM data. Experimental evaluation proves expected improvements in performance of watershed analysis with respect to a single-core CPU-based solution. It shows feasibility in using CUDA and OpenACC programming frameworks on GPUs and multi-core CPUs, for digital terrain analysis and similar GIS algorithms. Furthermore, our evaluation shows better performance of the OpenACC watershed analysis implementation with respect to the CUDA one, with gains in lower energy consumption and less programming efforts.

The main contributions of this paper are shown below.

- We have developed parallel implementation of all phases and steps of the watershed analysis using CUDA and OpenACC for NVIDIA GPUs and OpenACC for multi-core CPU.
- We have tested and evaluated OpenACC parallel implementation through several large DEM datasets both for multi-core CPUs and many-core GPUs, over commodity PC NVIDIA GPU, as well as high end NVIDIA Tesla K80 available through NVIDIA Docker plugin (*nvidia-docker*) through scientific cloud infrastructure.

- We have shown that the parallel implementation of OpenACC watershed analysis outperforms corresponding CUDA implementation for different DEM sizes, while consuming less energy and requiring less programming efforts.

The rest of the paper is organized as follows. Section 2 provides an overview of the related work in parallel and HPC implementation in DTA in general and watershed analysis in particular. Section 3 describes algorithms for watershed analysis, requirements, and motivation for their acceleration. Section 4 describes the parallel implementation of watershed algorithm, using CUDA and OpenACC programming frameworks and APIs (Application Programming Interfaces). Section 5 presents the experimental evaluation over multi-core CPU and many-core GPUs for large DEM datasets. Lastly, Section 6 concludes the paper and gives major directions for future research.

2. Related Work

Many GIS algorithms may benefit from parallel processing of geospatial data, such as: map-matching, viewshed analysis, watershed analysis, spatial overlay, trajectory analysis, and more [1]. The recent proliferation of distributed and cloud computing infrastructures and platforms, both public clouds (e.g., Amazon EC2) and private and hybrid computer clouds and clusters, has provided a rise in processing and analysis of geospatial data. The parallel GIS algorithms implemented on clusters of multi-core shared-memory computers, based on frameworks such as MPI (Message Passing Interface), MapReduce/Hadoop, and Apache Spark, have set this paradigm as an emerging research and development topic [3].

Zhang [4], Xia et al. [5], and Stojanovic and Stojanovic [6] have considered parallel processing of geospatial data in a personal computing environment. They argued that modern personal computers, equipped with multi-core CPUs and many-core GPUs, provide excellent support for parallel and high-performance processing of spatial data, which is comparable in efficiency and even in performance to cluster and cloud computing solutions using MPI or a MapReduce framework.

2.1. Remote Sensing and DTA Algorithms Implementation on a GPU

Many interesting and computationally and data-intensive algorithms in remote sensing and digital terrain analysis, such as flood modeling and simulations, LiDAR (Light Detection and Ranging) data processing, and watershed analysis have recently become the focus of research community oriented to parallel computing on many-core GPU architectures [7,8].

Li et al. [9] presented a parallel GPU implementation of map reprojection algorithms for raster datasets using CUDA. They demonstrated performance improvements over a serial version on a CPU through a series of tests, and achieved speedup from 10 to 100. Wang et al. [10] proposed a hybrid parallel spatial interpolation algorithm executed on both the CPU and GPU to increase the performance of massive LiDAR point clouds processing. Their experimental results have shown that their hybrid CPU-GPU solution outperforms the CPU-only and GPU-only implementations. Kang et al. [11] proposed a parallel algorithm implemented with OpenACC to use a GPU to parallelize the reservoir simulations. The experimental results show that the proposed approach outperform the CPU-based one while preserving the small programming effort for porting the algorithm to a parallel execution on a GPU. García-Feal et al. [12] presented a new parallel code in NVIDIA CUDA for two-dimensional (2D) flood inundation modelling using GPU. The experiments show a significant improvement in computational efficiency that opens up the possibility of using their solution for real-time forecasting of flood events as well. Liu et al. [13] proposed parallel OpenACC implementation of the two-dimensional shallow water model for flood simulation. The results of their experiments demonstrated achievements of up to one order of magnitude in speedup in comparison with the serial solution. Wu et al. [14] presented a parallel algorithm for DEM generalization implemented using CUDA applied in multi-scale terrain analysis. They proposed a parallel-multi-point algorithm for DEM generalization that outperforms other methods, such as ANUDEM (https://fennerschool.anu.edu.au/research/products/anudem), compound, and maximum z-tolerance method, while reducing the response time by up to 96%. Although not in focus in this research, OpenCL framework have been used in acceleration of several remote sensing and digital terrain analysis algorithms on GPU architectures as well [15,16]. OpenCL enables cross-platform parallel programming on heterogeneous platforms such as GPU, CPU, FPGA (Field-Programmable Gate Array), and DSP (Digital Signal Processing). Zhu et al. [15] have implemented a Non-Local means (NLM) denoising algorithm for image processing using OpenMP and OpenCL and conducted the experiment on CPU, GPU, and MIC (Many Integrated Core) Intel Xeon Phi Coprocessor. They show that OpenCL-based implementation has better performance on Xeon Phi 7120 than on NVIDIA Kepler K20M GPU, and slightly better performance than OpenMP-based implementation on Intel Xeon Phi 7120. Huang et al. [16] proposed implementation of a parallel compressive sampling matching pursuit (CoSaMP) for compressed sensing signal reconstruction using the OpenCL framework. Based on experiments using remote sensing images, they demonstrated that the proposed parallel OpenCL-based implementation can achieve significant speedup on heterogeneous platforms (AMD HD7350 and NVIDIA K20Xm), without modifying the application code. Since OpenCL and CUDA have similar parallel computing capabilities and expected performance improvements, we have not considered OpenCL in our implementation and experiments.

2.2. Watershed Analysis Implementation Using CUDA, OpenACC, and OpenCL

Different parallelization techniques and corresponding computer architectures have been particularly used for accelerating watershed analysis algorithms. Due to the recursive nature of the watershed algorithms, their parallelization is not a trivial task and it has been the focus of research for its acceleration through execution on parallel architectures with distributed memory [17–19], shared memory [20–22], and many-core GPUs [23–28]. Kauffmann and Piche [23] described a cellular automation to perform the watershed transformation in N-D images. Due to the local nature of cellular automation algorithms, these algorithms have been designed to execute on massively parallel processors and, therefore, be efficiently implemented on low cost consumer GPUs. Quesada-Barriuso et al. [24] showed that an algorithm for watershed analysis based on a cellular automaton is a good choice for implementing on the most recent GPU architectures, especially when the synchronization rules are relaxed. They compared the synchronous and asynchronous implementation of the algorithm. Their results showed high speedups for both implementations, especially for the asynchronous one.

Hučko and Šramek [25] presented a new algorithm for watershed analysis supporting data larger than the available memory. They modified one of the previous CPU intended algorithms for execution on a GPU architecture using OpenCL API. Two variants of the algorithm were designed and implemented. The tests showed no difference in the running time, which indicates that the storage of intermediate results in multi-pass algorithms consumes the main part of the running time and can be shortened using the RAID technology. It was shown that it was possible to reduce computational time of the watershed analysis approximately three times with respect to the original CPU version.

Ortega and Rueda were among the first to propose a GPU implementation of a classic watershed analysis algorithm [26]. They used CUDA for parallelizing the single-flow direction algorithm. They used a structure called the flow-transfer matrix to parallelize the D8 algorithm on a GPU. They achieved up to eight times speedup increase of CUDA-based drainage network computation with respect to the corresponding single-core implementation. Quin and Yhan [27] used CUDA to parallelize and accelerate both iterative DEM preprocessing step and a multiple-flow accumulation step of the watershed algorithm. Eranen et al. [28] implemented all the steps of the drainage network extraction algorithm on a GPU, for single flow directions, considering the uncertainty in the input digital elevation model.

Rueda et al. [29] implemented the flow accumulation step of the D8 algorithm in CPU, using OpenACC and two different CUDA versions, and compared the length and complexity of the programming code and its performance on different datasets. They concluded that, although OpenACC cannot match the performance of CUDA optimized implementation (3.5× slower in average), it provides a significant performance improvement against the CPU implementation (2–6×) with a much simpler parallel code and less implementation effort.

3. Algorithms for Watershed Analysis

Watersheds are catchment areas or drainage basins representing an extent from the land surface where water from different sources like rain, melting snow, and others, converges to the same point, called the outlet. Watersheds can be modeled considering a DEM as a grid model, where the cells represent the elevation of a square surface. Extracting a digital representation of the flow network is an essential step in the study of watershed delineation, erosion sites, mineral or pollution distribution, the cost estimation, the design of constructing new roads, the simulation of flood plains in paddy fields, and more.

The watershed analysis consists of two steps (phases): (1) DEM preprocessing concerning flow directions over flat areas and depressions in DEMs, and (2) Computation of flow distribution of each cell to its neighboring cells.

The DEM preprocessing algorithm is used to fill in depressions and remove flat areas existing in real DEMs. The result of depression filling in the DEM preprocessing phase is illustrated in Figure 1. The most proposed preprocessing algorithms are based on increasing the elevation of the cells located in a depression until the sink is filled (Figure 1) and the flow is routed over a flat area to the next lower cell. In this paper, the DEM preprocessing algorithm proposed in [30] is used.

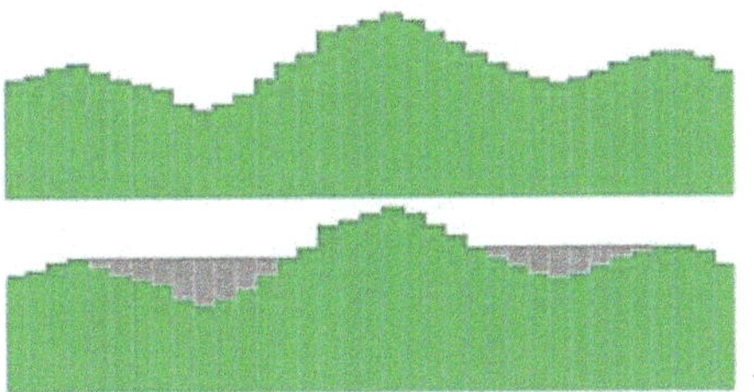

Figure 1. Depression filling.

The DEM preprocessing algorithm gets elevation raster zDEM as an input, and generates elevation raster wDEM as an output, by performing the following.

1. Adding a thick layer of water to each cell of elevation raster zDEM, except for the boundary cells.
2. Removing excess water since, for each cell, there is a non-increasing path to the boundary with two operations.

$$\text{zDEM}(c) \geq \text{wDEM}(n) + \varepsilon \Rightarrow \text{wDEM}(c) = \text{zDEM}(c) \tag{1}$$

$$\text{wDEM}(c) > \text{wDEM}(n) + \varepsilon \Rightarrow \text{wDEM}(c) = \text{wDEM}(n) + \varepsilon \tag{2}$$

where c represents a current cell on the position (i, j), n represents a boundary cell on the position (k, l), and ε is a slope of filled depression.

3. As long as the removal of the excess water (added in the first step) is possible, processing of the whole DEM is repeating iteratively in the while loop.

The second step in the watershed analysis represents computation of flow distribution from each cell to its neighboring cells. Two approaches can be used in this step: (i) Single Flow Direction (SFD) algorithms, where the flow is always passed to one of the eight neighboring cells, and (ii) Multiple Flow Direction (MFD) algorithms, where the flow is distributed to multiple neighbors, according to the predefined rule (Figure 2). In this paper, the MFD algorithm was used, which is better than SFD from the perspective of the algorithm error.

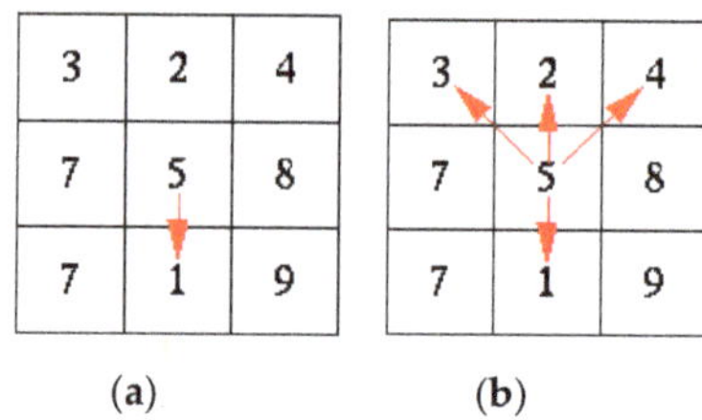

Figure 2. (**a**) Single flow direction. (**b**) Multiple flow direction.

In the MFD algorithm fraction of water d_i from the current cell to the i-th neighboring cell is calculated by the following expression.

$$d_i = \frac{(\tan \beta_i)^p \times L_i}{\sum_{j=1}^{8} (\tan \beta_j)^p \times L_j} \tag{3}$$

$$p = 8.9 \times \min(e, 1) + 1.1 \; e = \max \{ \tan \beta_1, \tan \beta_2, \ldots, \tan \beta_8 \}$$

where $\tan \beta_i$ represents the slope between the current cell and the *i*-th cell. The value of L_i depends on the position of the neighboring cell and it is calculated only for a cell whose elevation is less than the current cell elevation. Its value is 0.354 for diagonal cells and 0.5 for cardinal cells. In most MFD algorithms, the value of exponent p is equal to 1. There is a slight modification of the MFD algorithm, named MFD-md, where the value of p depends on terrain conditions, and it is calculated using Equation (3). Experimental results have shown that MFD-md gives more accurate results compared to classic MFD and SFD. Therefore, MFD-md was implemented. The MFD-md algorithm consists of two steps: (i) calculation of flow directions and (ii) calculation of flow accumulations. Calculation of flow directions is performed according to Equation (3), as shown in Figure 3.

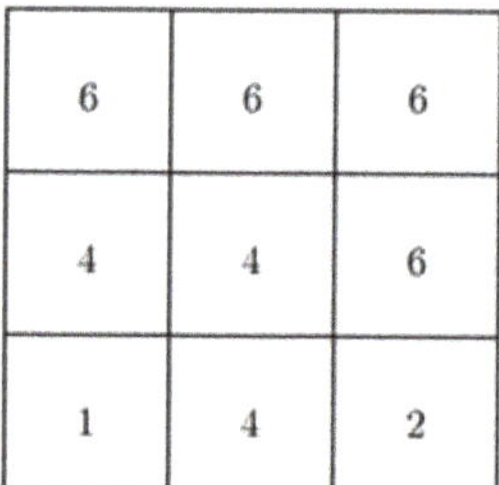
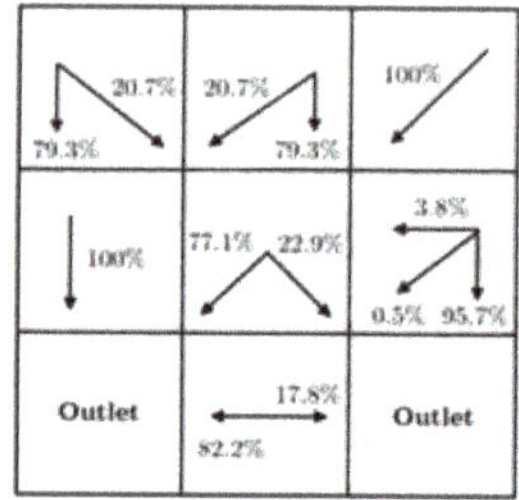

Figure 3. Calculation of flow directions in the MFD-md algorithm [18].

When flow directions are determined in DEM, in the second phase, the flow accumulations are calculated. In this paper, the Flow-Transfer Matrix (FTM)-based algorithm is used for parallelization of calculating the flow accumulation, as proposed in [26,27]. This is because its iterative version enables parallelization in each round of the iterative process. The FTM algorithm represents iterative process where, in each iteration (round), the flow accumulation for every cell, originated from its neighboring cells, is calculated. These values are recorded in a flow-transfer matrix, *FlowTransfer(i)*. The iterative process finishes when the zero flow-transfer matrix is obtained. The final *FlowAccumulation* matrix is obtained as the sum of *FlowTransfer(i)* matrices.

4. Parallelization of Watershed Analysis

In order to accelerate sequential execution of the watershed algorithm, we have implemented parallel applications using CUDA and OpenACC frameworks. Watershed analysis algorithm consists of two phases that can be executed either sequentially, or in parallel: (i) DEM preprocessing and (ii) computation of flow distribution of each cell to its neighboring cells. In this research, both steps in achieving a maximum increase in performance are parallelized. Both a CUDA solution for execution on a GPU and an OpenACC solution for execution on a GPU and a multi-core CPU have been implemented. The source code and executables for Watershed analysis implementation, as well as experimental DEM datasets, are available at public GitHub repository listed in Supplementary Materials section at the end of the paper.

As mentioned in Section 3, the DEM preprocessing algorithm in the first phase consists of two steps: the water covering step and the water removal step, and parallelization of both steps has been implemented. wDEM[i][j] > zDEM[i][j] is denoted as U(0). The conditions defined in Equations (1) and (2) are denoted as U(1) and U(2), respectively. The pseudo code for DEM preprocessing implementation is given in Figure 4.

```
1:   DEMpreprocessing (zDEM, wDEM, width, height, 0.01)
2:   {
3:       wDEM=WaterCovering(zDEM)
4:       finished=false
5:       while (!finished)
6:       {
7:           finished = true;
8:           for each cell [i][j] in DEM (except boundary, in any order)
9:           {
10:              if U(0) then
11:              {
12:                  for each neighbor [k][l] of cell [i][j] (in any order)
13:                  {
14:                      if U(1) then
15:                      {
16:                          operation (1); finished = false;
17:                      }
18:                      else if U(2) then
19:                      {
20:                          operation (2); finished = false;
21:                      }
22:                  }
23:              }
24:          }
25:      }
26: }
```

Figure 4. Pseudo code of the DEM preprocessing algorithm.

Regarding DEM preprocessing algorithm parallelization, we have used the fact that computation for the current cell is based on values of the neighboring cells that have been updated during the current iteration. As a consequence, computations in all cells can be performed in parallel during current iteration. This fact is exploited in both CUDA and OpenACC solutions.

As the pseudo-code inside double outer *for* loops (statements 13–22 in Figure 4) can be executed by single CUDA thread, its implementation can be placed into the appropriate kernel function. In this case, a double *for* loop is replaced with a kernel function. When the kernel function is called, a two-dimensional grid containing as many threads as the number of cells in the DEM is launched on the GPU. The kernel function is called from the host part of the algorithm (executed on a CPU) iteratively until none of the cells in DEM change their value during the current iteration of DEM preprocessing. Part of the host code is shown in Figure 5.

```
1:   while (!finished)
2:   {
3:       finished = true;
4:       d_finished ← finished;
5:       DepressionFillingKernel(dimGrid, dimBlock,
6:           d_wDEM,d_zDEM,width,height,epsilon, d_pFinished);
7:       finished ← d_finished;
8:   }
```

Figure 5. Part of the host pseudo code with the kernel launch.

In the case of OpenACC implementation, the independence of operations in the DEM processing algorithm is also taken into account. However, the parallelization is done in a different way in order to optimize execution of an OpenACC implementation. First of all, MFD algorithm is divided into functions: *DEMpreprocessing*, *GenerateFlowDirections*, *GenerateFlowFractions*, and *FlowAccumulation*. Each function contains OpenACC directives for parallel loop processing because there is no dependence of operations between loop iterations. In addition, it must be ensured that required data is stored in a GPU memory at the start of each parallel region. Data transfer has to be performed in an optimal way with respect to time consumption. For example, the input data for DEM preprocessing is zDEM and it must be transferred from the host to the device memory (GPU memory) beforehand. Since the DEM preprocessing function generates wDEM, it is not necessary to copy its values into the device memory. It is sufficient to allocate GPU memory for it. zDEM values are needed only during preprocessing, while wDEM is needed in other model processing functions. Thus, we need to assure that its values remain in the device memory during further execution of the successive functions.

One solution to the data transfer problem is to perform it once at the beginning of the parallel region in each of the four functions. In that case, the data transfer will be performed in each function, which is a time-consuming solution.

A better approach that was applied in this paper is to completely separate the OpenACC data transfer region and the parallel region. The necessary allocation of the device memory can be made and the necessary data can be copied from/to the host memory, using one OpenACC region (Figure 6).

```
#pragma acc data create(wDEM[:size], RMFD[:height][:width], flowFractions[:size1])
#pragma acc data copyin(zDEM[:size])
#pragma acc data copyout(flowAccumulation[:size])
{
    DEMpreprocessing(wDEM, zDEM, width, height, 0.01);
    GenerateFlowDirections(wDEM, RMFD, width, height);
    GenerateFlowFractions(wDEM, RMFD, flowFractions, cellSize, width, height);
    FlowAccumulation(wDEM, flowFractions, flowAccumulation, RMFD, width, height);
}
```

Figure 6. OpenACC parallel solution.

When calling any function involved in processing of the DEM model, the necessary data (wDEM) is available in the device memory. However, since each parallel region is out of the "scope" of the region that controls the parallel processing of data, the compiler cannot conclude that the data processed in each function has already been stored in the device memory. As a result, the compiler generates a generic data transfer region in each parallel region. To prevent this, we used additional OpenACC directives to notify the compiler that the data processed in a parallel region already exists in the device memory. Therefore, inside the DEM preprocessing function, before parallelization of both the water covering step and the water removal step, we added the following directive:

#pragma acc data present (wDEM [:size], zDEM [:size])

to determine compiler data processed in parallel regions that already exist in the device memory. The same directive has to be used before the parallel region in the *GenerateFlowDirections* function with wDEM [:size], RMFD [:height] [:width] as parameters, before the parallel region in *GenerateFlowFractions* function with wDEM[:size], RMFD[:height][:width], flowFractions [:size1] as parameters, and before the parallel region in *FlowAccumulation* function with wDEM [:length], flowFractions [:size1], RMFD [:height] [:width], flowAccumulation[:length] as parameters.

5. Experimental Evaluation

5.1. Experimental Settings

The proposed implementations have been tested on DEMs of different dimensions to evaluate efficiency and performance depending on the size of input data. Efficiency has been measured with respect to execution time, energy consumption, and programming efforts for algorithm parallelization. The experiment was carried out on the following parallel architectures.

- Intel(R) Core i5-4670K processor running at 3.40GHz, 8GB RAM (Random Access Memory), and NVIDIA GTX 770 graphic card.
- Tesla K80 GPU available on Hasso Plattner Institute (HPI) Future SOC (Service-Oriented Computing) Lab infrastructure (https://hpi.de/en/research/future-soc-lab-service-oriented-computing). NVIDIA Tesla K80 Accelerator contains 4992 cores, 24 GB of GDDR5 memory, and 480 GB/s memory bandwidth, according to the specification. We implemented the Docker image for our Watershed algorithm implementation and ran a container on Tesla K80 using NVIDIA Docker plugin (*nvidia-docker*) and its options.
- Intel® Xeon® Processor E5-2620 v4, with eight physical and 16 logical cores, 128 GB RAM, which is also available at HPI Future SOC Lab infrastructure.

We have applied the Watershed analysis over DEMs of several dimensions: 1691×2877, 2414×2912, 2433×4152, 3278×4277, and 3308×5967 to evaluate the scalability of implementation with the data size increasing. These DEMs represent the models of digital terrain in Alaska displayed in Figure 7 in the original form, using isohypses and *hill shade* relief, respectively, downloaded from the *EarthExplorer* USGS (United States Geological Survey) service (http://earthexplorer.usgs.gov).

To measure the execution time and the energy consumption for both GPU and CPU, we used the MeterPU library (https://www.ida.liu.se/labs/pelab/meterpu/). The execution times were measured in seconds and energy consumption in Joules. We also calculated programming productivity, i.e., programming effort for program parallelization Eff_{par} defined as:

$$Eff_{par} = 100 * \frac{LOC_{par}}{LOC_{tot}} \tag{4}$$

where LOC_{tot} (*Lines of Code total*) represents total lines of code and LOC_{par} (*Lines of Code parallel*) number of code lines, which are used for code parallelization.

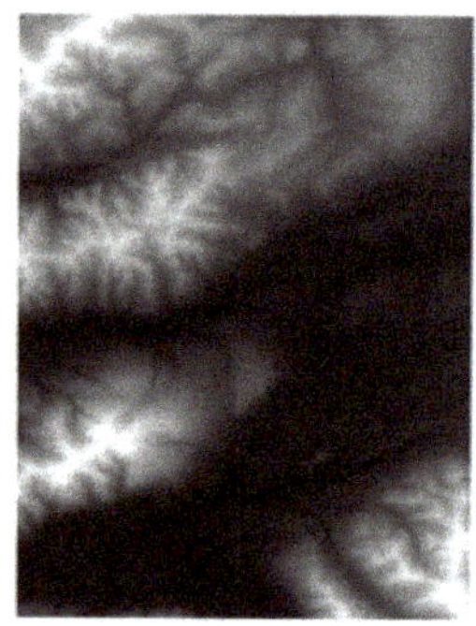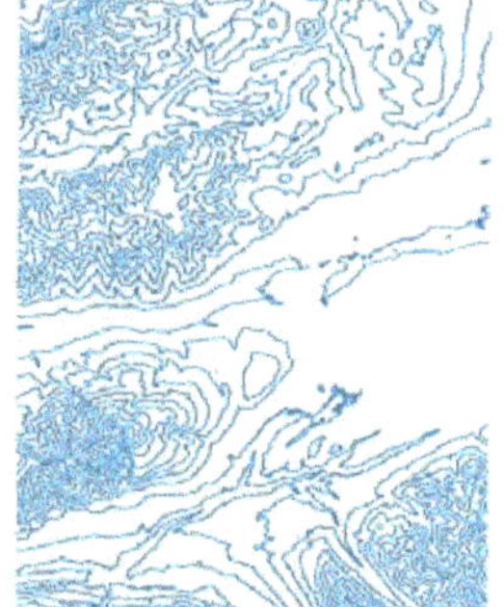

Figure 7. DEM for evaluation.

5.2. Experimental Results

For the implemented MFD algorithm, we have compared results for sequential (single-core) program execution, parallel multi-core CPU program execution (parallelized using OpenACC library), parallel many-core GPU execution (parallelized using OpenACC library), and parallel many-core GPU execution (parallelized using native CUDA code). The experimental results are shown in Tables 1–4.

According to these experiments, CPU and GPU execution times increases with DEM size, as expected. In the first dataset, CUDA and OpenACC GPU implementations outperform single-core CPU implementation by 11.2× and 25.7×, respectively, and multi-core CPU implementation by 5.4× and 12.5×, respectively. In the largest DEM dataset, CUDA and OpenACC GPU implementations outperform multi-core CPU implementation by 6.2× and 17.2×, respectively, and single-core CPU implementation by 12.3× and 34.2×, respectively.

Table 1. Experimental results for Single core (sequential) MFD.

Single Core (Sequential) MFD		
DEM Size	**Execution Time (s)**	**Programming Effort for Program Parallelization**
1691 × 2827	126.11	
2414 × 2912	253.65	
2433 × 4152	370.63	0
3278 × 4227	632.03	
3308 × 5967	866.43	

Table 2. Experimental results for Multicore-OpenACC MFD.

Multi-Core OpenACC MFD		
DEM Size	**Execution Time (s)**	**Programming Effort for Program Parallelization**
1691 × 2827	61.33	
2414 × 2912	119.65	
2433 × 4152	176.91	0.03
3278 × 4227	304.57	
3308 × 5967	434.89	

Table 3. Experimental results for CUDA many-core GPU MFD.

CUDA Many-Core GPU MFD		
DEM Size	**Execution Time (s)**	**Programming Effort for Program Parallelization**
1691 × 2827	11.30	
2414 × 2912	19.85	
2433 × 4152	29.35	0.31
3278 × 4227	48.11	
3308 × 5967	70.53	

Table 4. Experimental results for OpenACC many-core GPU MFD.

OpenACC Many-Core GPU MFD		
DEM Size	**Execution Time (s)**	**Programming Effort for Program Parallelization**
1691 × 2827	4.89	
2414 × 2912	8.35	
2433 × 4152	12.57	0.03
3278 × 4227	18.79	
3308 × 5967	25.35	

For better performance evaluation, we calculated the Speedup (S) to compare the execution time of parallel implementations of the MFD algorithm ($T_{parallel}$) to the execution time of sequential MFD implementation on a single core CPU processor ($T_{sequential}$):

$$S = \frac{T_{seqential}}{T_{parallel}} \tag{5}$$

Figure 8 shows experimental results for the speedup of the corresponding parallel solution with respect to the sequential solution for different sizes of input DEM.

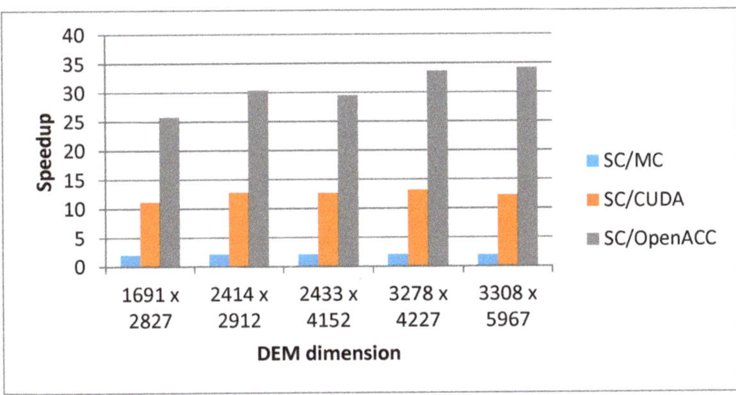

Figure 8. Speedup for the MFD algorithm.

It can be concluded that all parallel implementations have speedup greater than 1 with respect to the sequential solution (marked as SC in Figure 8). Thus, OpenACC multi-core implementation (marked as MC in Figure 8) shows the speedup is, on average, 2.1, CUDA many-core implementation (marked as CUDA in Figure 8) shows the speedup, on average, is 12.4, and OpenACC many-core implementation (marked as OpenACC in Figure 8) shows the speedup, on average, is 30.1. The results in Figure 8 illustrate that OpenACC many-core GPU implementation achieves higher speedup than counterpart CUDA implementations. The reason is found in the fact that, in the OpenACC implementation, all phases of DEM processing through to the final result are parallelized: preprocessing, generating auxiliary matrices (a matrix that stores the direction of the water flow and a matrix that memorizes the percentage of water swelling in each direction), while, in the CUDA implementation, only the preprocessing and processing of DEM data are parallelized. In addition, the OpenACC implementation eliminates unnecessary data transfers between the host and GPU at the beginning of the execution of each phase, which are performed in the CUDA version. This has a significant impact on the overall execution time of OpenACC implementation.

Since there are no parallel constructions in the serial implementation, the programming effort for the serial implementation is zero. There are no instructions for parallel processing (Table 1). Since the same code was extended with the OpenACC parallel processing directives used to generate multi-core programs and many-core GPU programs, the value of the programming effort is the same for both, which is 0.03 (Tables 2 and 4). This value is less than 1%, or less than 1% of the instruction of the entire code referring to the parallelization of the program. On the other hand, the programming effort for the original CUDA application is 0.3012 (Table 3), or almost one third of the instructions in the code are instructions for parallel processing. As is shown, the optimal solution is based on OpenACC applied on a GPU because it is easy to parallelize to obtain satisfactory results.

In order to measure the execution time and the energy consumption of Watershed analysis, the MeterPU library is provided for advanced architectures, such as NVIDIA Tesla K80 and Intel® Xeon® Processor E5-2620 v4 CPU that we used at a Future SOC Lab. The experimental results of the sequential and parallel CUDA and OpenACC implementations on these architectures are presented in Tables 5–8. The Speedup achieved for different parallel implementations with respect to the sequential ones is presented in Table 9 and displayed in Figure 9.

Table 5. Single core-Intel® Xeon® Processor E5-2620 v4.

Single core—Intel® Xeon® Processor E5-2620 v4			
DEM Size	Execution Time (s)	Energy Consumption—CPU (J)	Energy Consumption—GPU (J)
1691 × 2827	159.776	12,135.8	4871.3
2414 × 2912	315.515	24,538.2	9619.4
2433 × 4152	459.171	35,769.8	14,007.4
3278 × 4227	781.827	61,060.6	23,862.2
3308 × 5967	1090.010	85,613.0	33,238.4

Table 6. Multi-core—OpenACC-Intel® Xeon® Processor E5-2620 v4.

Multi-core—OpenACC-Intel® Xeon® Processor E5-2620 v4			
DEM Size	Execution Time (s)	Energy Consumption—CPU (J)	Energy Consumption—GPU (J)
1691 × 2827	9.5034	1307.65	290.28
2414 × 2912	17.0548	2469.39	519.83
2433 × 4152	37.1679	5228.46	1132.68
3278 × 4227	43.5649	6579.67	1328.08
3308 × 5967	61.8061	9400.93	1883.37

Table 7. OpenACC-Tesla K80.

	OpenACC-Tesla K80		
DEM Size	**Execution Time (s)**	**Energy Consumption—CPU (J)**	**Energy Consumption—GPU (J)**
1691 × 2827	6.2446	435.93	690.64
2414 × 2912	10.0951	707.68	1147.29
2433 × 4152	14.2851	942.79	1697.09
3278 × 4227	19.8665	1469.80	2442.86
3308 × 5967	26.8563	1973.31	3359.60

Table 8. CUDA—native-Tesla K80.

	CUDA—Native-Tesla K80		
DEM Size	**Execution Time (s)**	**Energy Consumption—CPU (J)**	**Energy Consumption—GPU(J)**
1691 × 2827	9.7792	714.52	1099.04
2414 × 2912	16.6563	1196.26	1945.94
2433 × 4152	24.0191	1737.13	2849.46
3278 × 4227	39.0244	2875.51	4789.63
3308 × 5967	55.9863	4104.54	6941.10

Table 9. Speedup (Tesla K80).

	Speedup (Tesla K80)		
DEM	**SC/MC**	**SC/CUDA**	**SC/OpenACC**
1691 × 2827	16.8125	16.3383	25.5863
2414 × 2912	18.5001	18.9427	31.2543
2433 × 4152	12.3540	19.1169	32.1434
3278 × 4227	17.9463	20.0343	39.3540
3308 × 5967	17.6360	19.4692	40.5868

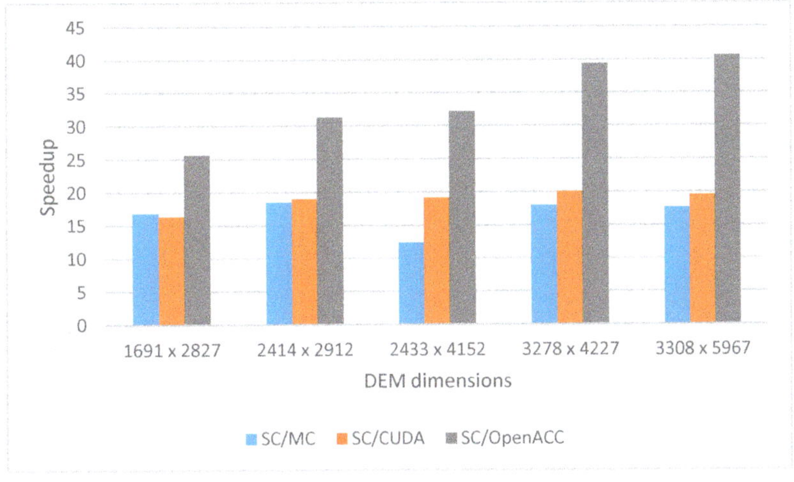

Figure 9. Speedup Tesla K80.

Based on Tables 5–9 and Figure 9, it can be concluded that OpenACC multi-core CPU implementation (marked as MC in Figure 9) shows speedup on average 16.6 (maximum 18.5), CUDA many-core GPU implementation (marked as CUDA in Figure 9) shows speedup on average 18.7 (maximum 20), and OpenACC many-core GPU implementation shows speedup on average 33.7 (maximum 40.6). These results show similar behavior to the previous results shown in Figure 8, and they confirm previously presented conclusions. The obtained results are better with respect to the results shown in Figure 8 because, this time, we used more powerful hardware. The significant difference in Speedup is especially visible in the case of OpenACC multi-core CPU solution since the Intel Xeon processor is used.

Regarding energy consumption, all parallel solutions consume less total energy (CPU+GPU) than the corresponding sequential solution. The total energy consumption of multi-core OpenACC solution is comparable to the CUDA GPU solution. The first one consumes more CPU energy while the second one consumes more GPU energy (Figures 10 and 11). When comparing GPU-based solutions, OpenACC implementation consumes almost two times less energy than its CUDA counterpart (Figure 11).

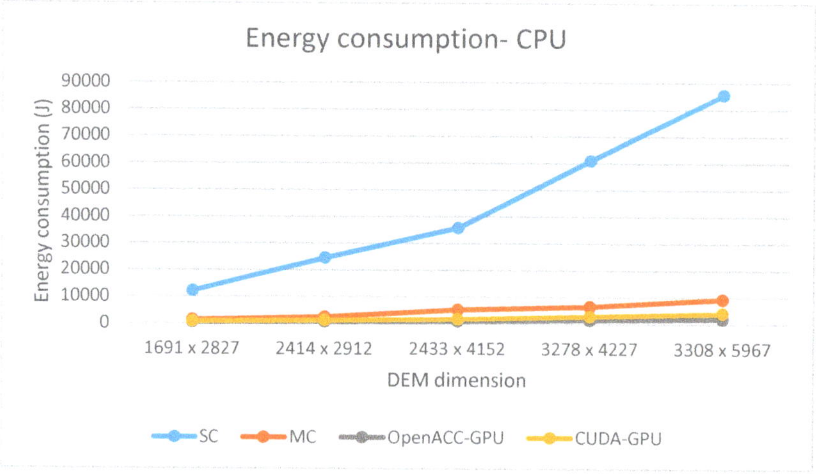

Figure 10. Energy consumption CPU-Intel Xeon E5-2620 v4 @ 2.10GHz.

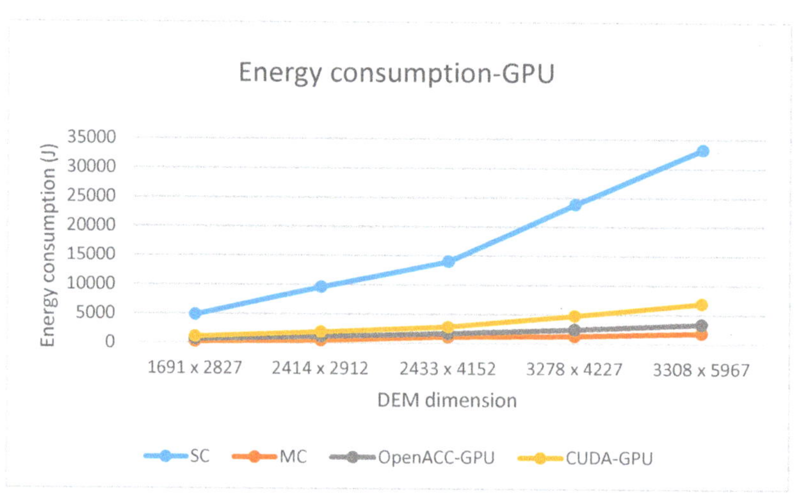

Figure 11. Energy consumption GPU-Tesla K80.

6. Conclusions

Advances in remote sensing, geo-sensor networks, and mobile positioning in recent years, have provided a generation of massive geospatial data of various formats. This has caused great interest in processing, analysis, and visualization of Big geospatial and spatio-temporal data, both offline and as fast data streams in recent years. GPGPU represents a new parallel computing paradigm that provides significant gains in term of performance improvements in various data intensive applications. This paper shows that using OpenACC and CUDA parallel programming paradigms can significantly improve the performance in executing various computation and data intensive GIS algorithms and that using parallelization and high-performance computing in GIS represents a promising direction for research and development.

The benefits of parallel processing of geospatial data is confirmed in parallelization of watershed analysis algorithms and they are evaluated on multi-core CPU and many-core GPU using CUDA and OpenACC frameworks.

Adaptation of sequential watershed algorithm implementation to many-core GPU requires significant code transformations and optimizations for CUDA parallel implementation, and that is why we considered OpenACC for parallel implementation for many-core GPU, but also for multi-core CPU. OpenACC requires less development effort, lower risk of errors, and better code readability with respect to the CUDA GPU solution. We have implemented the watershed analysis algorithm that consists of two computationally intensive and time-consuming phases: (1) iterative DEM preprocessing and (2) iterative MFD algorithm, which are both suitable for parallelization. Experimental evaluation indicates improvement in performance with respect to a single-core CPU-based solution and shows feasibility of using GPU and multicore CPU for watershed analysis. The evaluation of proposed solutions is performed with respect to execution time, energy consumption, and programming effort for program parallelization for a different size of input data. The experimental results show benefits of using the OpenACC framework over CUDA for parallelization of watershed analysis using the MFD algorithm and, thus, its feasibility for GIS analytic algorithms over Big raster and vector geospatial data.

Supplementary Materials: The source code and executables for Watershed analysis implementation, as well as experimental DEM datasets are available online at https://github.com/drstojanovic/WatershedAnalysis.

Author Contributions: Conceptualization, N.S. and D.S. Methodology, N.S. and D.S. Investigation, N.S. and D.S. Software, N.S. and D.S. Validation, N.S. and D.S. Formal Analysis, N.S. and D.S. Writing-Original Draft Preparation, N.S. and D.S. Writing-Review & Editing, N.S. and D.S. Supervision, N.S. and D.S.

Funding: The Ministry of Education, Science and Technological Development, Republic of Serbia, as part of the project "Environmental Protection and Climate Change Monitoring and Adaptation", III-43007, partly funded this research.

Acknowledgments: The Ministry of Education, Science and Technological Development, Republic of Serbia, as part of the project "Environmental Protection and Climate Change Monitoring and Adaptation," III-43007 and Research grant for Hasso-Plattner-Institute (HPI) Future SOC Lab cloud computing infrastructure (https://hpi.de/en/research/future-soc-lab-service-oriented-computing.html) supported research presented in this paper.

Conflicts of Interest: The authors declare no conflict of interest.

References

1. Stojanovic, N.; Stojanovic, D. High-performance computing in GIS: Techniques and applications. *Int. J. Reason. Based Intell. Syst. IJRIS* **2013**, *5*, 42–49. [CrossRef]

2. Kirk, D.; Hwu, W.M. *Programming Massively Parallel Processors: A Hands-on Approach*; Elsevier: Amsterdam, The Netherlands, 2010.

3. Stojanovic, N.; Stojanovic, D. A hybrid MPI + OpenMP application for processing big trajectory data. *Stud. Inform. Control* **2015**, *24*, 229–236. [CrossRef]

4. Zhang, J. Towards personal high-performance geospatial computing (HPC-G): Perspectives and a case study. In Proceedings of the ACM SIGSPATIAL—HPDGIS 2010 Workshop, San Jose, CA, USA, 2–5 November 2010; pp. 3–10. [CrossRef]

5. Xia, Y.; Li, Y.; Shi, X. Parallel viewshed analysis on GPU using CUDA. In Proceedings of the 3rd International Joint Conference on Computational Science and Optimization, Huangshan, China, 28–31 May 2010; Volume 1, pp. 373–374. [CrossRef]

6. Stojanovic, N.; Stojanovic, D. High performance processing and analysis of geospatial data using CUDA on GPU. *Adv. Electr. Comput. Eng.* **2014**, *14*, 109–114. [CrossRef]

7. Strnad, D. Parallel terrain visibility calculation on the graphics processing unit. *Concurr. Comput. Pract. Exp.* **2011**, *23*, 2452–2462. [CrossRef]

8. Stojanovic, N.; Stojanovic, D. Performance improvement of viewshed analysis using GPU. In Proceedings of the 11th International Conference on Telecommunications in Modern Satellite, Cable and Broadcasting Services (TELSIKS), Nis, Serbia, 16–19 October 2013; pp. 397–400. [CrossRef]

9. Li, J.; Finn, M.P.; Castano, M.B. A lightweight CUDA-based parallel map reprojection method for raster datasets of continental to global extent. *ISPRS Int. J. Geo Inf.* **2017**, *6*, 92. [CrossRef]

10. Wang, H.; Guan, X.; Wu, H. A hybrid parallel spatial interpolation algorithm for massive LiDAR point clouds on heterogeneous CPU-GPU systems. *ISPRS Int. J. Geo Inf.* **2017**, *6*, 363. [CrossRef]

11. Kang, Z.; Deng, Z.; Han, W.; Zhang, D. Parallel reservoir simulation with OpenACC and domain decomposition. *Algorithms* **2018**, *11*, 213. [CrossRef]

12. García-Feal, O.; González-Cao, J.; Gómez-Gesteira, M.; Cea, L.; Manuel Domínguez, J.; Formella, A. An accelerated tool for flood modelling based on Iber. *Water* **2018**, *10*, 1459. [CrossRef]

13. Liu, Q.; Qin, Y.; Li, G. Fast simulation of large-scale floods based on GPU parallel computing. *Water* **2018**, *10*, 589. [CrossRef]

14. Wu, Q.; Chen, Y.; Wilson, J.P.; Liu, X.; Li, H. An effective parallelization algorithm for DEM generalization based on CUDA. *Environ. Model. Softw.* **2019**, *114*, 64–74. [CrossRef]

15. Zhu, H.; Wu, Y.; Li, P.; Wang, D.; Shi, W.; Zhang, P.; Jiao, L. A parallel Non-Local means denoising algorithm implementation with OpenMP and OpenCL on Intel Xeon Phi Coprocessor. *J. Comput. Sci.* **2016**, *17*, 591–598. [CrossRef]

16. Huang, F.; Tao, J.; Xiang, Y.; Liu, P.; Dong, L.; Wang, L. Parallel compressive sampling matching pursuit algorithm for compressed sensing signal reconstruction with OpenCL. *J. Syst. Archit.* **2017**, *72*, 51–60. [CrossRef]

17. Plaza, A.; Plaza, J.; Valencia, D.; Martinez, P. Parallel segmentation of multi-channel images using multi-dimensional mathematical morphology. In *Advances in Image and Video Segmentation*; IGI Global: Hershey, PA, USA, 2006; pp. 270–291. [CrossRef]

18. Wu, S.; Yingshuai, H. Parallelization research on watershed algorithm. In Proceedings of the International Conference on Automatic Control and Artificial Intelligence (ACAI), Xiamen, China, 24–26 March 2012; pp. 1524–1527. [CrossRef]

19. Świercz, M.; Iwanowski, M. Fast, parallel watershed algorithm based on path tracing. In Proceedings of the International Conference on Computer Vision and Graphics, Warsaw, Poland, 20–22 September 2010; Springer: Berlin, Germany, 2010; pp. 317–324. [CrossRef]

20. Wagner, B.; Dinges, A.; Müller, P.; Haase, G. Parallel volume image segmentation with watershed transformation. In Proceedings of the Scandinavian Conference on Image Analysis, Oslo, Norway, 15–18 June 2009; Lecture Notes in Computer Science 5575. Springer: Berlin/Heidelberg, Germany, 2009; pp. 420–429. [CrossRef]

21. Mahmoudi, R.; Akil, M. Real-time topological image smoothing on shared memory parallel machines. In Proceedings of the Real-Time Image and Video Processing, San Francisco, CA, USA, 24–25 January 2011; Proc.SPIE 7871. p. 787109. [CrossRef]

22. Van Neerbos, J.; Najman, L.; Wilkinson, M.H.F. Towards a parallel topological watershed: First results. In Proceedings of the International Symposium on Mathematical Morphology and Its Applications to Signal and Image Processing, Verbania-Intra, Italy, 6–8 July 2011; Springer: Berlin/Heidelberg, Germany, 2011; pp. 248–259. [CrossRef]

23. Kauffmann, C.; Piche, N. Cellular automaton for ultra-fast watershed transform on GPU. In Proceedings of the 19th International Conference on Pattern Recognition, Tampa, FL, USA, 8–11 December 2008; IEEE: New York, NY, USA, 2008; pp. 1–4. [CrossRef]

24. Quesada-Barriuso, P.; Heras, D.B.; Argüello, F. Efficient GPU asynchronous implementation of a watershed algorithm based on cellular automata. In Proceedings of the 10th IEEE International Symposium on Parallel and Distributed Processing with Applications, Leganés, Spain, 10–13 July 2012; IEEE: New York, NY, USA, 2012; pp. 79–86. [CrossRef]
25. Hučko, M.; Šrámek, M. Streamed watershed transform on GPU for processing of large volume data. In Proceedings of the 28th Spring Conference on Computer Graphics, Budmerice, Slovakia, 2–4 May 2012; ACM: New York, NY, USA, 2013; pp. 137–141. [CrossRef]
26. Rueda, L.; Ortega, A.J. Parallel drainage network computation on CUDA. *Comput. Geosci.* **2010**, *36*, 171–178. [CrossRef]
27. Qin, C.Z.; Zhan, L. Parallelizing flow accumulation calculations on graphics processing units from iterative DEM preprocessing algorithm to recursive multiple-flow direction algorithm. *Comput. Geosci.* **2012**, *43*, 7–16. [CrossRef]
28. Eränen, D.; Oksanen, J.; Westerholm, J.; Sarjakoski, T. A full graphics processing unit implementation of uncertainty-aware drainage basin delineation. *Comput. Geosci.* **2014**, *73*, 48–60. [CrossRef]
29. Rueda, A.J.; Noguera, J.M.; Luque, A. A comparison of native GPU computing versus OpenACC for implementing flow-routing algorithms in hydrological applications. *Comput. Geosci.* **2016**, *87*, 91–100. [CrossRef]
30. Planchon, O.; Darboux, F. A fast, simple and versatile algorithm to fill the depressions of digital elevation models. *CATENA* **2002**, *46*, 159–176. [CrossRef]

International Journal of
Geo-Information

Article

LandQ^{v2}: A MapReduce-Based System for Processing Arable Land Quality Big Data

Xiaochuang Yao [1,*] , **Mohamed F. Mokbel** [2], **Sijing Ye** [3], **Guoqing Li** [1], **Louai Alarabi** [2], **Ahmed Eldawy** [4], **Zuliang Zhao** [5] , **Long Zhao** [5] **and Dehai Zhu** [5,*]

1   Institute of Remote Sensing and Digital Earth, Chinese Academy of Sciences, Beijing 100094, China; ligq@radi.ac.cn
2   Department of Computer Science and Engineering, University of Minnesota, Minneapolis, MN 55455, USA; mokbel@umn.edu (M.F.M.); louai@cs.umn.edu (L.A.)
3   State Key Laboratory of Earth Surface Processes and Resource Ecology, Beijing Normal University, Beijing 100875, China; yesj@bnu.edu.cn
4   Department of Computer Science and Engineering, University of California, Riverside, CA 92521, USA; eldawy@cs.ucr.edu
5   College of Information and Electrical Engineering, China Agricultural University, Beijing 100083, China; zlzhao@cau.edu.cn (Z.Z.); zhaolcau@163.com (L.Z.)
*   Correspondence: yaoxc@radi.ac.cn (X.Y.); zhudehai@263.net (D.Z.); Tel.: +86-1881-135-6282 (X.Y.)

Received: 25 May 2018; Accepted: 6 July 2018; Published: 10 July 2018

Abstract: Arable land quality (ALQ) data are a foundational resource for national food security. With the rapid development of spatial information technologies, the annual acquisition and update of ALQ data covering the country have become more accurate and faster. ALQ data are mainly vector-based spatial big data in the ESRI (Environmental Systems Research Institute) shapefile format. Although the shapefile is the most common GIS vector data format, unfortunately, the usage of ALQ data is very constrained due to its massive size and the limited capabilities of traditional applications. To tackle the above issues, this paper introduces LandQ^{v2}, which is a MapReduce-based parallel processing system for ALQ big data. The core content of LandQ^{v2} is composed of four key technologies including data preprocessing, the distributed R-tree index, the spatial range query, and the map tile pyramid model-based visualization. According to the functions in LandQ^{v2}, firstly, ALQ big data are transformed by a MapReduce-based parallel algorithm from the ESRI Shapefile format to the GeoCSV file format in HDFS (Hadoop Distributed File System), and then, the spatial coding-based partition and R-tree index are executed for the spatial range query operation. In addition, the visualization of ALQ big data with a GIS (Geographic Information System) web API (Application Programming Interface) uses the MapReduce program to generate a single image or pyramid tiles for big data display. Finally, a set of experiments running on a live system deployed on a cluster of machines shows the efficiency and scalability of the proposed system. All of these functions supported by LandQ^{v2} are integrated into SpatialHadoop, and it is also able to efficiently support any other distributed spatial big data systems.

Keywords: spatial big data; parallel processing; MapReduce; arable land quality (ALQ); GIS

---

1. Introduction

Arable land is a foundational resource for national food security. Arable land quality (ALQ) data are about the quality evaluation of arable land by classification and gradation. With the rapid development of spatial information technologies, the annual acquisition and update of ALQ data covering the country have become more accurate and faster. ALQ data are dominated by spatial vector data in the ESRI shapefile format. In China, the government began to perform all round arable

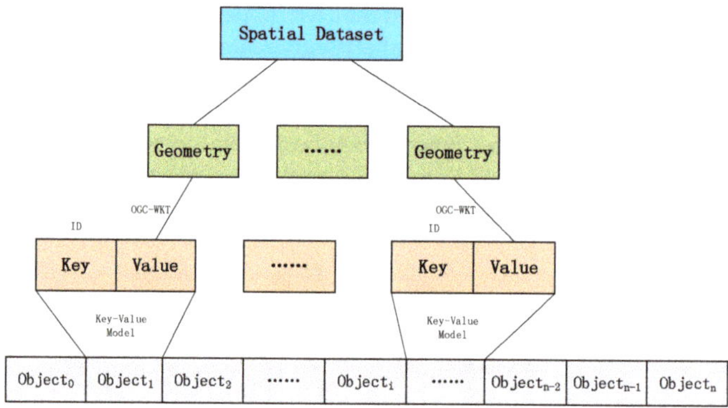

Figure 2. The spatial vector data cloud storage model, GeoCSV [18].

Based on the GeoCSV model, this model has the following advantages: (1) object-oriented storage of existing spatial vector data, and the parallel computation of fine granularity is supported; (2) simple data structure with OGC standard data and the advantages of network transmission; (3) CSV is used for the separation of storage, which is conducive to the distributed processing of data segmentation, processing, and analysis; and (4) spatial data and attribute data formats are unified, and it is easy to add a field or adjust the data structure.

3.1.2. Data Conversion

Data conversion from Shapefile to GeoCSV is executed by one MapReduce job as shown in Figure 3. In this way, batch conversions can be implemented, which will change multiple small Shapefiles into one large GeoCSV file. Since a Shapefile is composed of multiple sub-files (.shp, .shx, .dbf, etc.), and each sub-file contains a unique header file, the conversion of a single Shapefile file will be treated as a separate subtask in the implementation of the data conversion algorithm. The task is done primarily as a two-part process, namely, Shapefile parsing and spatial object reconstruction in GeoCSV.

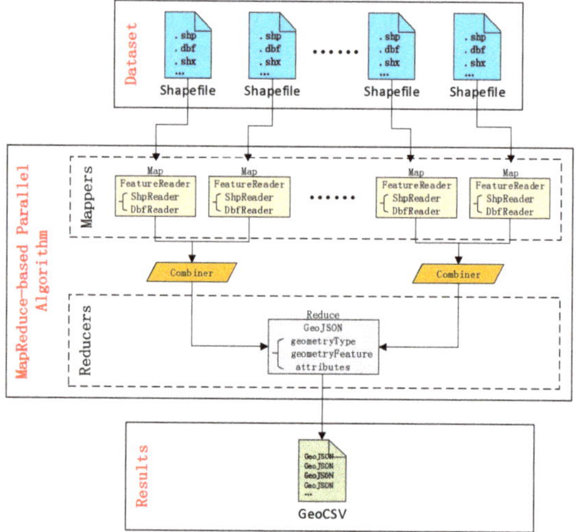

Figure 3. The MapReduce-based parallel data conversion algorithm.

3.2. Spatial Index

In the distributed storage environment, a spatial dataset will be separated into several blocks dispensed to different nodes in the cluster [20], and the spatial index technology based on a single node cannot be used directly. Compared with the centralized index, the distributed index increases the communication protocol and communication overhead among the nodes in the cluster [21]. The index mechanism of a distributed storage system must take full account of the distributed systems and the organization of the data storage.

3.2.1. Data Partitioning

In this section, we present the spatial partitioning architecture based on the spatial coding-based approach (SCA) [22]. Figure 4 summarizes the six steps, which can be done by two MapReduce jobs. (1) Computing the spatial location: the central point is used instead of the spatial objects; (2) Defining the spatial coding matrix (SCM): we define the spatial coding values as a spatial coding matrix; (3) Computing the sensing information set (SIS): the SIS includes the spatial code, location, size, and count; (4) Computing the spatial partitioning matrix (SPM): SPM is built according to the SIS and the blocked size in HDFS; (5) Spatial data partitioning: every spatial object will be matched with the spatial coding and written into the data blocks in HDFS; (6) Allocating the data blocks: all blocks will be distributed into the nodes in the cluster with their spatial codes.

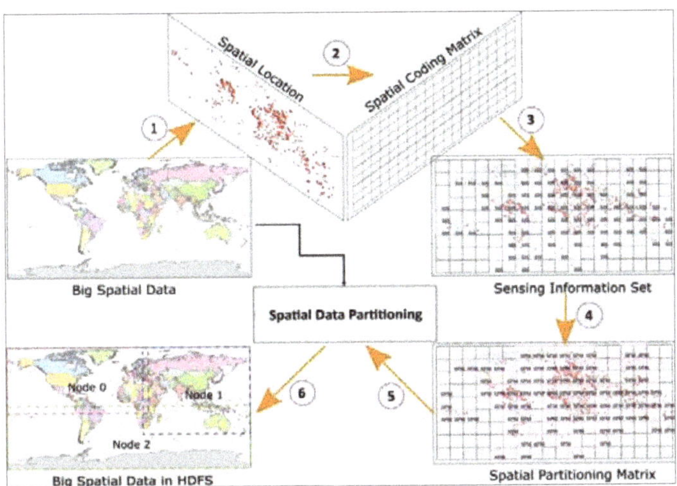

Figure 4. The algorithm of spatial partitioning based on SCA (Spatial Coding-based Approach) [22].

3.2.2. Distributed R-Tree Index

The R-tree index is the most influential data access method in the GIS database [23]. As Figure 5 shows, the distributed R-tree index includes the local index and global index [24]. The local index, also called a data block index, is created for the collection of spatial object information in the data block after the spatial partitioning steps in HDFS. Here, this index will generate an index header file for each data block in HDFS, and the contents of the index header file include the envelope rectangle (MBR), the count of spatial objects (Count), offset to the start (Offset), and other basic information. The global index is created from the local index. The purpose of this phase is to build the global index structure, indexing all partitions in HDFS. By concatenating all the local index files into one file, the global index file will be generated. In addition, the content of the global index file includes the data block id (ID), the envelope rectangle (MBR), the total size (Size), and others.

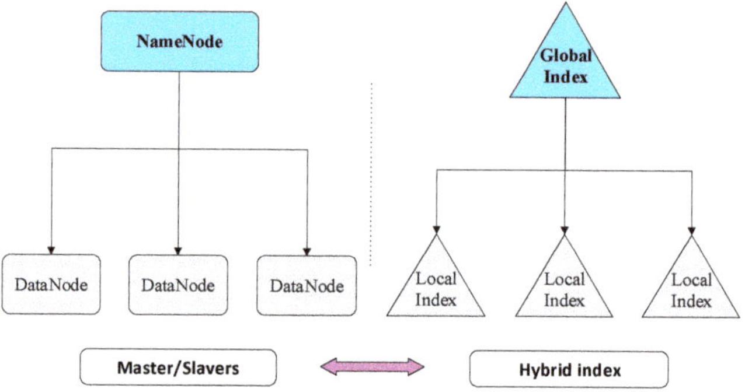

Figure 5. The distributed R-tree index.

3.3. Spatial Range Query

The spatial query operation is closely related to the spatial data storage model and spatial index, and it can be regarded as their reverse process. In the non-indexed Hadoop cloud environment, a spatial query needs to traverse all the spatial information records to match the corresponding results, and for large-scale spatial data, the query efficiency is extremely low.

As shown in Figure 6, this paper combines the spatial range query algorithm into three phases: (a) the Global filtering phase. The content stored in the global index is the basic information of all the data blocks in HDFS. The size of the global index file is relatively small, so this phase is executed at the master node. All query operations need to go through the global filtering phase first and the global index file usually resides in the memory of the master node; (b) the Local filtering stage. The local filtering phase is mainly for filtering the candidate sets in the data blocks from the global filtering results. The local index file is mainly the leaf node of the distributed R-tree stored in the data block header file; (c) the Refining stage. Through step (b), it can get all the leaf nodes in the current data block that intersects the query range. The refining stage mainly matches each spatial geometric object in the candidate set; if the object intersects or is included in the spatial query range, it will be put in the results file, otherwise, it will not perform this operation.

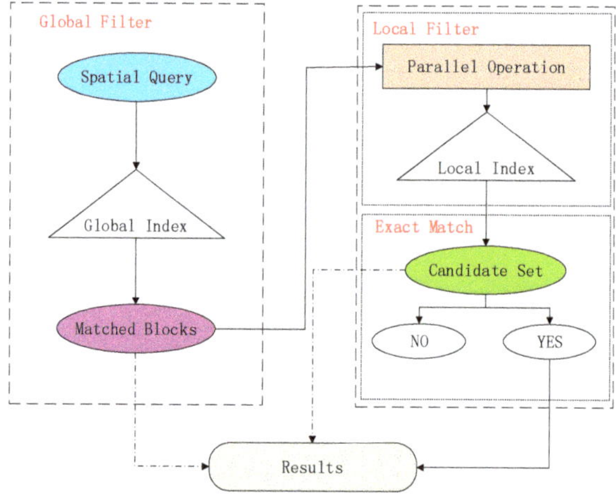

Figure 6. The spatial query steps based on the distributed R-tree index.

3.4. Data Visualization

The tile pyramid model uses "vertical grading, horizontal block" thinking, and the map data are divided into uniform rectangular tiles where the number of levels depends on the map scale, and the tile number is decided by the picture size [25].

The visualization based on the tile pyramid model is developed by employing a three phase technique [26]: (1) The partitioning phase: here, we use the default Hadoop partitioning to split the total input spatial big data into partitions in HDFS; (2) The plotting phase: through the internal loop, all tiles occupied by each spatial object will be plotted in the tile pyramid model; (3) The merging phase: the partial images having the same index code will be combined into one final image. As shown in Figure 7, each tile in the map pyramid model is indexed according to the level, row, and column as the unique identifier. In the parallel algorithm, the construction of each tile can be regarded as a separate sub-task, and the entire tile construction of the pyramid model can be regarded as a collection of multiple subtasks.

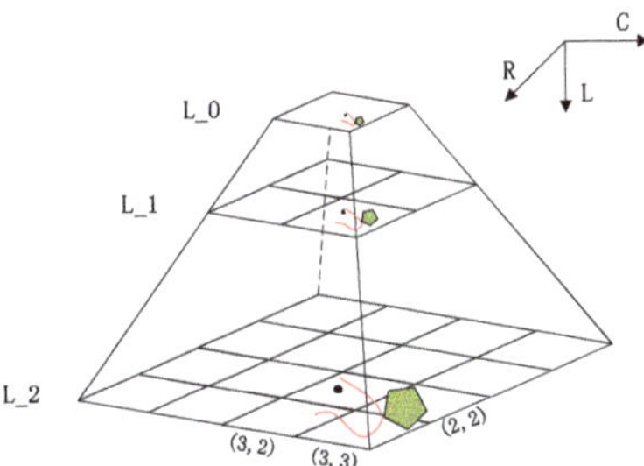

Figure 7. The schematic diagram of the tile pyramid model.

In general, we assume that the enclosing rectangle of the space object shape is mbr, and the maximum and minimum coordinates are $(x1, y1)$ and $(x2, y2)$, respectively. Then, the following formula is used to get the spatial object in the pyramid model. The cover of the map tile (r) column (c) is

$$r_{min}(\text{shape}) = \left\lceil \frac{mbr_{y1} - MBR_{y1}}{MBR_{length}/2^l} \right\rceil \tag{1}$$

$$r_{max}(\text{shape}) = \left\lceil \frac{mbr_{y2} - MBR_{y1}}{MBR_{length}/2^l} \right\rceil \tag{2}$$

$$c_{min}(\text{shape}) = \left\lceil \frac{mbr_{x1} - MBR_{x1}}{MBR_{width}/2^l} \right\rceil \tag{3}$$

$$c_{max}(\text{shape}) = \left\lceil \frac{mbr_{x2} - MBR_{x1}}{MBR_{width}/2^l} \right\rceil \tag{4}$$

where the $[r_{min}, r_{max}]$ and $[c_{min}, c_{max}]$ for the space object shape of the map tile correspond to the line number and column number interval, respectively, and MBR is for the entire map. The l is the pyramid level.

In particular, if the map tile size is p and the level has a spatial resolution of r, the spatial resolution of the tile in the x and y directions satisfies the following conditions:

$$r_x = \frac{MBR_{width}}{p \times 2^l} \tag{5}$$

$$r_y = \frac{MBR_{length}}{p \times 2^l} \tag{6}$$

4. System Test and Results

In this section, the key technologies mentioned above are tested with the national arable land quality (ALQ) big data. The content of the test in this paper includes three aspects: data conversion, spatial query operation, and visualization performance. The test results will be given bellow.

4.1. Data Conversion

The parallel conversion experiment of the vector data mainly includes two aspects. First, the parallel algorithm proposed in this paper is compared with the data transformation algorithm in the ArcGIS software. On the other hand, the parallel algorithm is tested in the cluster environment. The experimental environment in this paper includes the stand-alone and cluster environment. Table 1 configures the information for the experimental environment. Table 2 shows the test data and that the size of the ESRI Shapefile varies from 2 GB to 128 GB.

Table 1. The test environment for the data conversion algorithm.

Test Environment	System Configuration
Stand-alone	Windows 10, 64 GB Memory, 600 GB Hard disk ArcGIS 10.2 Ubuntu 15, 64 GB Memory, 600 GB Hard disk Hadoop 2.7
Cluster	Ubuntu 15, 24 GB Memory, 2 TB Hard disk Hadoop 2.7

Table 2. The datasets for the parallel conversion test.

Size/GB	Shapefile Number	Feature Number
2	86	1,361,127
4	257	2,512,413
8	304	5,218,028
16	680	11,077,580
32	1120	21,615,328
64	3008	43,847,424
128	5561	85,558,229

The experimental results are shown in Figures 8 and 9. It can be seen that the conversion time of the two algorithms increases with the growth of the size of the ESRI Shapefile. At the same time, it is fairly clear that the parallel conversion algorithm based on the MapReduce proposed in this paper is more efficient in terms of time. Moreover, with the increase of the number of ESRI Shapefile, the efficiency growth is clearer, approximately 4 to 6 times to the ArcToolBox.

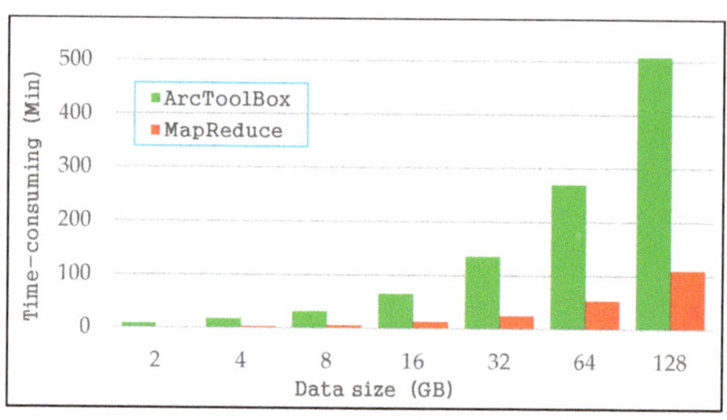

Figure 8. The time-consuming comparison of data conversion in a stand-alone environment.

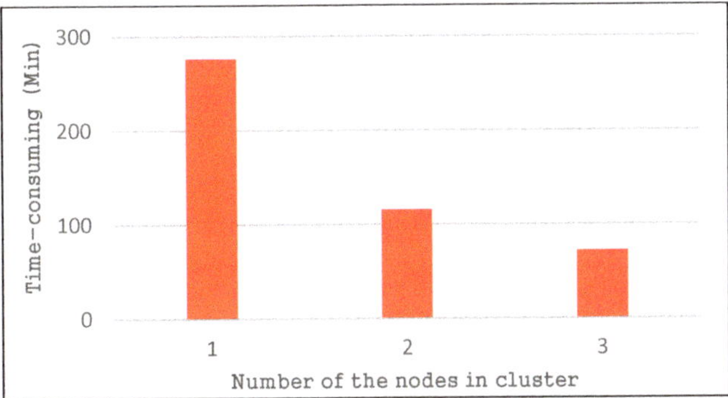

Figure 9. The time-consuming comparison of data conversion in a cluster environment.

Figure 9 shows the time consuming comparison of the data conversion efficiency in the cluster environment. Here, we only test a data volume of 64 GB. Through the test results, we can find that the execution time of the algorithm becomes shorter with the increase of the number of cluster nodes, which fully embodies the advantages of the cloud computing environment.

4.2. Data Query

4.2.1. Index Quality of the R-Tree

The R-tree spatial index quality can be measured by the envelope rectangle of its index tree hierarchies. The calculation formula is as follows, where Area (T) in Equation (7) represents the total area covered by the MBR of each index layer of the R-tree. Generally, the less the coverage area is, the better the R-tree index performance. In Equation (8), Overlap (T) represents the overlap area between the MBRs of each index layer in the R-tree. The larger the overlap area is, the more space the query path will have, which leads to a greater query time consumption and greatly reduces the efficiency of the data retrieval. Thus, for overlapping areas, the smaller the area is, the better the R-tree performance will be. The R-tree spatial index quality analysis formula is as follows:

$$\text{Area}(T) = \sum_{i=1}^{n}(T_i MBR) \tag{7}$$

$$\text{Overlap}(T) = \sum_{i=1}^{n}\sum_{j=i+1}^{n} Area\left(T_i MBR \cap T_j MBR\right) \tag{8}$$

As shown in Figure 10, spatial coding-based data partition (Hilbert coding-based approach, HCA) shows a better index quality than random sampling for building an R-tree in both Area (T) and Overlay (T). The main reasons include the following two aspects. (1) Due to the randomness of the sample set and lost spatial features. The data partitioning method based spatial-coding can make good use of the adjacent features of spatial coding, which greatly guarantees the spatial distribution of the vector data; (2) the data partitioning method based on spatial sampling. Only the spatial location information of the sample is considered, however, the other one is considered with more influencing factors, including spatial location information, the number of elements, the amount of data bytes, and so on. In addition, the sampling based method is also due to the randomness of the samples, the partitioning strategy for the same dataset and the same sampling rate are different, and the algorithm lacks stability.

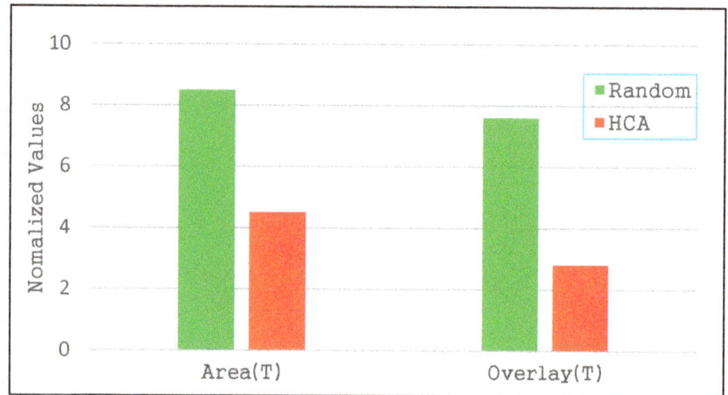

Figure 10. The index quality comparison of the R-tree based on random and HCA.

4.2.2. Spatial Query Performance

Based on the distributed R-tree, here, we test the query performance of LandQ^{v2}. We measure the query time consumption in the different numbers of jobs and nodes.

In experiment 1, the result in Figure 11 shows the query performance comparison between the different job numbers with two modes, namely, the non-index and R-tree index. It can be seen that the spatial query execution time for the data increases with the increase of the spatial parallel access times, and there is a linear trend of growth for the non-index model, which is mainly due to traversing all the data blocks regardless of any spatial query. For the R-tree index model, because the search is only executed with the relevant data block, there is no obvious linear growth trend and there is a great relationship with the spatial query itself. In addition, for both models, the more query jobs are queried, the longer it will take.

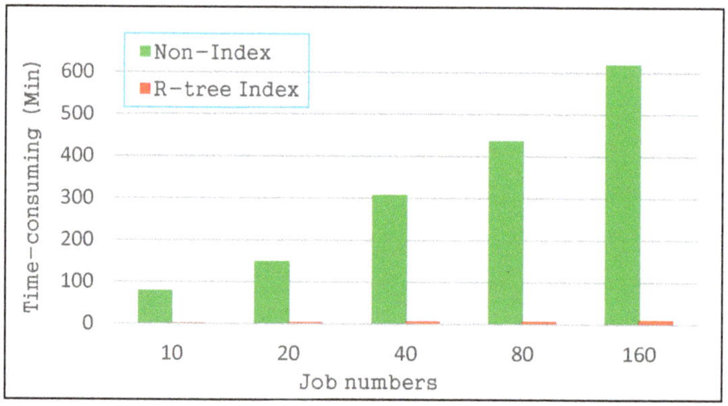

Figure 11. The query performance comparison between the different job numbers.

Experiment 2 tests the vector data query efficiencies with spatial index in different cluster nodes. The experimental results are shown in Figure 12. It can be seen that the efficiency of the data query increases with the increase in the number of cluster nodes. At the same time, the parallel query algorithm in this paper can perform well and shows good superiority.

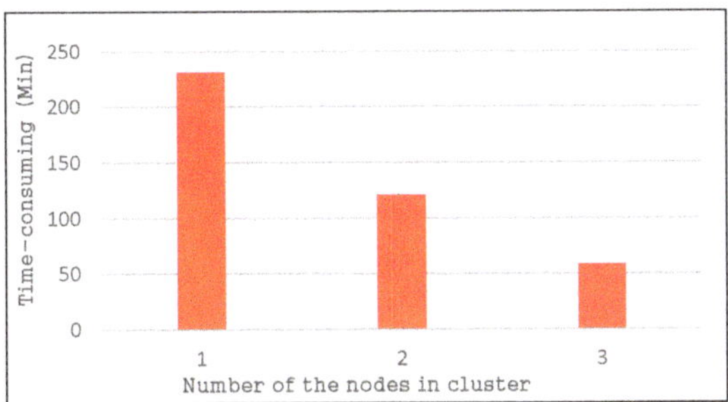

Figure 12. The query performance comparison between the different node numbers.

Through the above comparison tests, the following conclusions can be drawn. (1) The efficiency of the spatial query based on the spatial index has been obviously improved; (2) The R-tree index based on the spatial coding-data partitioning also showed a good advantage in the spatial query operation; (3) With the increase of the nodes, the efficiency of the spatial parallel query can be significantly improved, and it can expand the number of clusters to improve the efficiency of the data retrieval.

4.3. Visualization Performance

4.3.1. Tile Pyramid Construction Performance

In this section, we test two experiments. One experiment is a comparison between the parallel construction algorithm proposed in this paper and the existing mainstream ArcGIS server. The other tests the algorithm performance in the cluster. The experimental data is in the Table 3.

Table 3. The datasets for visualization.

Dataset	Size	Shapefile Count	Feature Count
Province	3.4 GB	31	359,227
County	154.5 GB	2673	63,033,494

Experiment 1 was conducted in the same hardware environment and the tile pyramid model parameters were consistent. The test results are shown in Figure 13. Under the same test environment, the algorithm proposed in this paper is better. In experiment 2, the efficiency of the parallel construction of different datasets is compared with the number of different cluster nodes. The tile pyramid level is set to level 8. The result is shown in Figure 14. Under the same task, the greater the number of cluster nodes was, the clearer the advantages of tile pyramid parallel construction were.

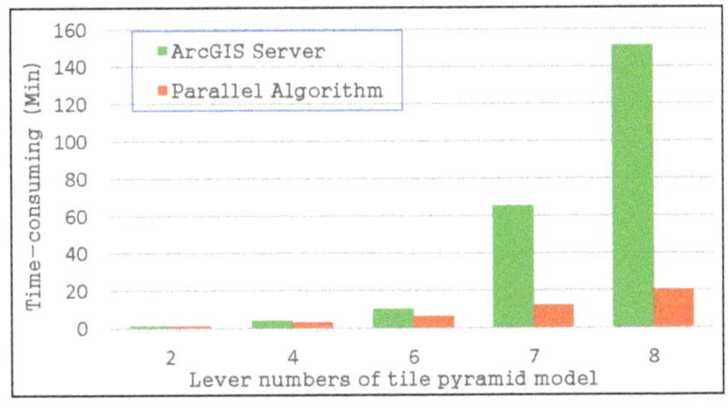

Figure 13. The time-consuming comparison of the tile pyramid construction in a single machine environment.

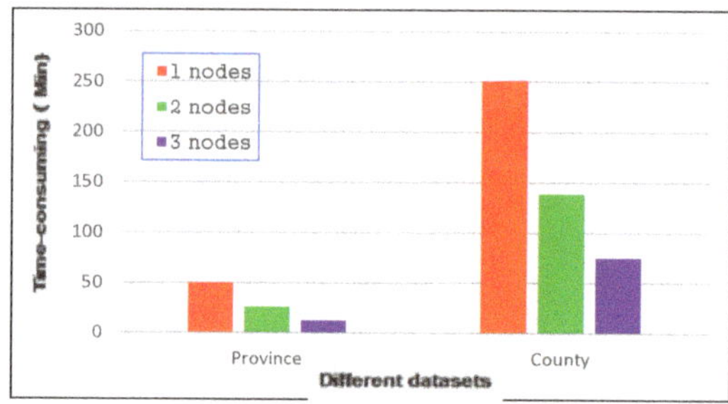

Figure 14. The time-consuming comparison of the tile pyramid construction in a cluster environment.

4.3.2. Visualization for Arable Land Quality Big Data

Based on the parallel construction algorithm of the tile pyramid model, the provincial and county level of arable land quality (ALQ) big data are respectively visualized. Following the OGC map standard, we construct the map tiles and integrate them with the ESRI API for JavaScript very well.

For the 2013 national and provincial level of arable land quality, there were equal data map elements of the tile pyramid parallel construction. In this paper, we showed the national economic classification, which reflects the economic value of the arable land. The resulting tile details are shown in Table 4. The tile pyramid model has 8 levels of map scale ranging from 1:20 million to 1:31.25 million. The county dataset had a parallel build time of 1 h 15 min 7 s, and the provincial data had a parallel build time of 11 min and 52 s.

Table 4. The datasets for the parallel conversion test.

Level	Scale (million)	Tile Number (County)	Size (County)	Tile Number (Province)	Size (Province)
0	1:4000	6	35.1 KB	6	36.4 KB
1	1:2000	15	103 KB	15	100 KB
2	1:1000	44	331 KB	44	504 KB
3	1:500	124	1.06 MB	123	837 KB
4	1:250	390	3.71 MB	380	2.41 MB
5	1:125	1312	12.3 MB	1265	6.6 MB
6	1:62.5	4523	36.9 MB	4311	16.6 MB
7	1:31.25	15,138	111 MB	15,183	40.1 MB

Figures 15 and 16 show the national arable land quality data global browsing and local browsing diagrams, respectively. In line with the WMS standard, the resulting map tiles are stored in folders with different scales which will be accessed and loaded by the Web client. At the same time, it can integrate well with other layers, such as national administrative boundary, railway, water, and other layers in the client.

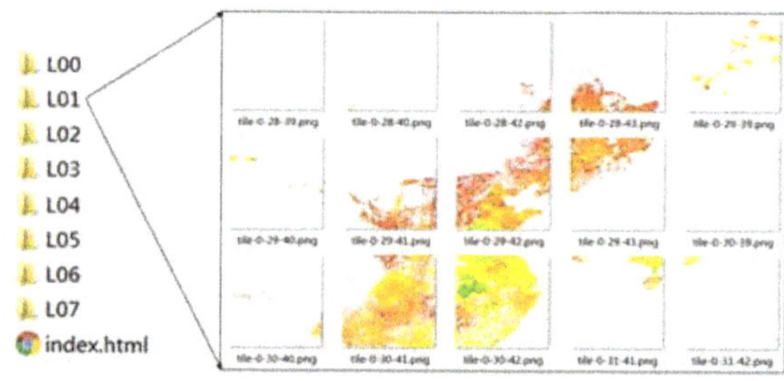

Figure 15. The map tiles of the national arable land quality dataset.

Figure 16. The schematic diagram of visualization for the national arable land quality dataset.

5. Discussion

The increased amount of the arable land quality (ALQ) data has actually brought challenges for processing and managing the vector-based spatial big data, especially in terms of data processing efficiency. Traditional GIS solutions have been unable to meet the needs of the national application. Here, we present a MapReduce-based parallel processing system, LandQ^{v2}, for national arable land quality (ALQ) big data. LandQ^{v2} is a complete set of application systems and is integrated into SpatialHadoop. Similarly, the relevant technologies can also be applied to other areas.

In LandQ^{v2}, four key technologies are designed and implemented. It is based on the cloud storage model of the spatial vector data (GeoCSV). The many shortcomings of the ESRI Shapefile will be addressed and solved in the data storage and process. The batch conversion algorithm with MapReduce improves the efficiency of data processing from the ESRI Shapefile to GeoCSV. Aimed at the problem of data query efficiency, a spatial coding-based approach for partitioning the spatial big data is presented, and then the distributed R-tree index is built for the spatial range query. This method will improve the query performance of the spatial big data, as well as the data balance in HDFS [22]. For the visualization of the remote sensing data, the tile pyramid model is a good and mature solution [27,28]. In this paper, we use this model to solve the problem of visualizing the vector-based spatial big data. The tile pyramid model-building algorithm is proposed and parallelized with the MapReduce program.

We tested the above key technologies and system functions. By comparing these features with existing GIS tools, all of the parallel algorithms, including data conversion, data partitioning, distributed R-tree index, data query, and visualization, show good performance and, at the same time, they can take full advantages of the cluster scalability. Compared with LandQ^{v1}, the visualization of the ALQ dataset in national level makes the macroscopic grasp of data more comprehensive and scientific, which will be useful to policymakers.

6. Conclusions

Recent advancements in cloud computing technology have offered a potential solution to the big data challenges [29]. This paper presents LandQ^{v2}, which is a MapReduce-based parallel processing system for arable land quality (ALQ) big data that uses the Hadoop cloud computing platform. The contents of system architecture, key technologies, and the tests shown in this study are efficient enough to bring the following remarkable advantages of the developed system: (1) The spatial vector big data cloud storage model, GeoCSV, which is based on the characteristics of spatial vector data and the advantages of the Hadoop cloud platform; (2) the distributed R-tree index in LandQ^{v2}, the data partitioning strategy based on spatial coding is designed, and the parallel construction of the distributed R-tree index is realized. Through experiments, the efficiency and feasibility of different distributed spatial indexing algorithms are verified from two perspectives: spatial index quality and data balance; (3) parallel processing for ALQ big data. This paper carries out parallel processing methods including a data conversion algorithm, spatial range query, and tile pyramid model construction. All these parallel algorithms are implemented with the MapReduce program. In addition, using ALQ big data, the experiments are used to verify the efficiency and feasibility of the proposed vector data processing algorithm.

Author Contributions: X.Y. developed the LandQ^{v2} and wrote this paper. M.F.M., L.A. and A.E. are the contributors of the SpatialHadoop and provided key technical guidance for the system development. S.Y., Z.Z., and L.Z. assisted in the system development. G.L. and D.Z. are the supporters of this project and supervised this paper.

Funding: This work was sponsored by National Key Research and Development Program of China from MOST (2016YFB0501503).

Acknowledgments: We would like to thank the anonymous reviewers and editors for commenting on this paper.

Conflicts of Interest: The authors declare no conflict of interest.

References

1. Yao, X.; Zhu, D.; Ye, S.; Yun, W.; Zhang, N.; Li, L. A field survey system for land consolidation based on 3S and speech recognition technology. *Comput. Electron. Agric.* **2016**, *127*, 659–668. [CrossRef]
2. Ye, S.; Zhu, D.; Yao, X.; Zhang, N.; Fang, S.; Li, L. Development of a highly flexible mobile GIS-based system for collecting arable land quality data. *IEEE J. Sel. Top. Appl. Earth Obs. Remote Sens.* **2014**, *7*, 4432–4441. [CrossRef]
3. Yao, X.; Yang, J.; Li, L.; Yun, W.; Zhao, Z.; Ye, S.; Zhu, D. LandQ^{v1}: A GIS cluster-based management information system for arable land quality big data. In Proceedings of the 6th International Conference on Agro-Geoinformatics (Agro-Geoinformatics), Fairfax, VA, USA, 7–10 August 2017; pp. 1–6.
4. Huang, Q.Y.; Yang, C.W.; Liu, K.; Xia, J.Z.; Xu, C.; Li, J.; Gui, Z.P.; Sun, M.; Li, Z.L. Evaluating open-source cloud computing solutions for geosciences. *Comput. Geosci.* **2013**, *59*, 41–52. [CrossRef]
5. Li, Z.; Yang, C.; Liu, K.; Hu, F.; Jin, B. Automatic scaling Hadoop in the cloud for efficient process of big geospatial data. *ISPRS Int. Geo-Inf.* **2016**, *5*, 173. [CrossRef]
6. Aji, A.; Sun, X.; Vo, H.; Liu, Q.; Lee, R.; Zhang, X.; Saltz, J.; Wang, F. Demonstration of Hadoop-GIS: A spatial data warehousing system over mapreduce. In Proceedings of the 21st ACM SIGSPATIAL International Conference on Advances in Geographic Information Systems, Orlando, FL, USA, 5–8 November 2013; ACM: New York, NY, USA, 2013; pp. 528–531.
7. Eldawy, A.; Mokbel, M.F. A demonstration of spatialhadoop: An efficient mapreduce framework for spatial data. *Proc. VLDB Endow.* **2013**, *6*, 1230–1233. [CrossRef]
8. Hughes, J.N.; Annex, A.; Eichelberger, C.N.; Fox, A.; Hulbert, A.; Ronquest, M. Geomesa: A distributed architecture for spatio-temporal fusion. In Proceedings of the Geospatial Informatics, Fusion, and Motion Video Analytics V, Baltimore, MD, USA, 20–21 April 2015; p. 94730F.
9. Yu, J.; Wu, J.; Sarwat, M. Geospark: A cluster computing framework for processing large-scale spatial data. In Proceedings of the 23rd ACM SIGSPATIAL International Conference on Advances in Geographic Information Systems, Seattle, WA, USA, 3–6 November 2015; ACM: New York, NY, USA, 2015; pp. 1–4.
10. Alarabi, L. St-Hadoop: A mapreduce framework for big spatio-temporal data. In Proceedings of the ACM International Conference on Management of Data, Chicago, IL, USA, 14–19 May 2017; ACM: New York, NY, USA, 2017; pp. 40–42.
11. Mueller, N.; Lewis, A.; Roberts, D.; Ring, S.; Melrose, R.; Sixsmith, J.; Lymburner, L.; McIntyre, A.; Tan, P.; Curnow, S.; et al. Water observations from space: Mapping surface water from 25 years of landsat imagery across Australia. *Remote Sens. Environ.* **2016**, *174*, 341–352. [CrossRef]
12. Li, J.; Meng, L.; Wang, F.Z.; Zhang, W.; Cai, Y. A map-reduce-enabled solap cube for large-scale remotely sensed data aggregation. *Comput. Geosci.* **2014**, *70*, 110–119. [CrossRef]

13. Zhong, Y.; Fang, J.; Zhao, X. Vegaindexer: A distributed composite index scheme for big spatio-temporal sensor data on cloud. In Proceedings of the 33rd IEEE International Geoscience and Remote Sensing Symposium, IGARSS, Melbourne, VIC, Australia, 21–26 July 2013; pp. 1713–1716.

14. Magdy, A.; Mokbel, M.F.; Elnikety, S.; Nath, S.; He, Y. Venus: Scalable real-time spatial queries on microblogs with adaptive load shedding. *IEEE Trans. Knowl. Data Eng.* **2016**, *28*, 356–370. [CrossRef]

15. Addair, T.G.; Dodge, D.A.; Walter, W.R.; Ruppert, S.D. Large-scale seismic signal analysis with Hadoop. *Comput. Geosci.* **2014**, *66*, 145–154. [CrossRef]

16. Zou, Z.Q.; Wang, Y.; Cao, K.; Qu, T.S.; Wang, Z.M. Semantic overlay network for large-scale spatial information indexing. *Comput. Geosci.* **2013**, *57*, 208–217. [CrossRef]

17. Jhummarwala, A.; Mazin, A.; Potdar, M.B. Geospatial Hadoop (GS-Hadoop) an efficient mapreduce based engine for distributed processing of shapefiles. In Proceedings of the the 2nd International Conference on Advances in Computing, Communication, & Automation, Bareilly, India, 30 September–1 October 2016; pp. 1–7.

18. Yao, X.; Li, G. Big spatial vector data management: A review. *Big Earth Data* **2018**, *2*, 108–129. [CrossRef]

19. OGC. Geographic Information-Well-Known Text Representation of Coordinate Reference Systems. Available online: http://docs.opengeospatial.org/is/12-063r5/12-063r5.html (accessed on 20 June 2018).

20. Zhao, L.; Chen, L.; Ranjan, R.; Choo, K.-K.R.; He, J. Geographical information system parallelization for spatial big data processing: A review. *Clust. Comput.* **2015**, *19*, 139–152. [CrossRef]

21. Singh, H.; Bawa, S. A mapreduce-based scalable discovery and indexing of structured big data. *Future Gener. Comput. Syst.* **2017**, *73*, 32–43. [CrossRef]

22. Yao, X.; Mokbel, M.F.; Alarabi, L.; Eldawy, A.; Yang, J.; Yun, W.; Li, L.; Ye, S.; Zhu, D. Spatial coding-based approach for partitioning big spatial data in Hadoop. *Comput. Geosci.* **2017**, *106*, 60–67. [CrossRef]

23. Hadjieleftheriou, M.; Manolopoulos, Y.; Theodoridis, Y.; Tsotras, V.J. R-trees—A dynamic index structure for spatial searching. In *Encyclopedia of GIS*; Shekhar, S., Xiong, H., Eds.; Springer: Boston, MA, USA, 2008; pp. 993–1002.

24. Eldawy, A.; Alarabi, L.; Mokbel, M.F. Spatial partitioning techniques in spatialhadoop. *Proc. VLDB Endow.* **2015**, *8*, 1602–1605. [CrossRef]

25. Zhang, J.; You, S. High-performance quadtree constructions on large-scale geospatial rasters using GPGPU parallel primitives. *Int. J. Geogr. Inf. Sci.* **2013**, *27*, 2207–2226. [CrossRef]

26. Eldawy, A.; Mokbel, M.F.; Jonathan, C. Hadoopviz: A mapreduce framework for extensible visualization of big spatial data. In Proceedings of the 32nd IEEE International Conference on Data Engineering, Helsinki, Finland, 16–20 May 2016; pp. 601–612.

27. Liu, Y.; Chen, L.; Jing, N.; Xiong, W. Parallel batch-building remote sensing images tile pyramid with mapreduce. *Wuhan Daxue Xuebao (Xinxi Kexue Ban)/Geomat. Inf. Sci. Wuhan Univ.* **2013**, *38*, 278–282.

28. Lin, W.; Zhou, H.; Xia, P. An effective NOSQL-based vector map tile management approach. *ISPRS Int. Geo-Inf.* **2016**, *5*, 1–25.

29. Lee, J.-G.; Kang, M. Geospatial big data: Challenges and opportunities. *Big Data Res.* **2015**, *2*, 74–81. [CrossRef]

International Journal of
Geo-Information

Article

Mr4Soil: A MapReduce-Based Framework Integrated with GIS for Soil Erosion Modelling

Zhigang Han [1,2,3,*], **Fen Qin [1,2,*]**, **Caihui Cui [1,3]**, **Yannan Liu [4]**, **Lingling Wang [5]** and **Pinde Fu [6]**

[1] College of Environment and Planning, Henan University, Kaifeng 475004, China; chcui@henu.edu.cn
[2] Key Laboratory of Geospatial Technology for the Middle and Lower Yellow River Regions, Ministry of Education, Kaifeng 475004, China
[3] Urban Big Data Institute, Henan University, Kaifeng 475004, China
[4] School of Computer and Communication Engineering, Zhengzhou University of Light Industry, Zhengzhou 450002, China; lyn2018@zzuli.edu.cn
[5] Yellow River Institute of Hydraulic Research, Yellow River Conservancy Commission of the Ministry of Water Resources, Zhengzhou 450003, China; wanglingling@hky.yrcc.gov.cn
[6] Environmental Systems Research Institute, Inc. Redlands, 380 New York Street, Redlands, CA 92373-8100, USA; pfu@esri.com
* Correspondence: zghan@henu.edu.cn (Z.H.); qinfen@henu.edu.cn (F.Q.)

Received: 10 January 2019; Accepted: 22 February 2019; Published: 27 February 2019

Abstract: A soil erosion model is used to evaluate the conditions of soil erosion and guide agricultural production. Recently, high spatial resolution data have been collected in new ways, such as three-dimensional laser scanning, providing the foundation for refined soil erosion modelling. However, serial computing cannot fully meet the computational requirements of massive data sets. Therefore, it is necessary to perform soil erosion modelling under a parallel computing framework. This paper focuses on a parallel computing framework for soil erosion modelling based on the Hadoop platform. The framework includes three layers: the methodology, algorithm, and application layers. In the methodology layer, two types of parallel strategies for data splitting are defined as row-oriented and sub-basin-oriented methods. The algorithms for six parallel calculation operators for local, focal and zonal computing tasks are designed in detail. These operators can be called to calculate the model factors and perform model calculations. We defined the key-value data structure of GeoCSV format for vector, row-based and cell-based rasters as the inputs for the algorithms. A geoprocessing toolbox is developed and integrated with the geographic information system (GIS) platform in the application layer. The performance of the framework is examined by taking the Gushanchuan basin as an example. The results show that the framework can perform calculations involving large data sets with high computational efficiency and GIS integration. This approach is easy to extend and use and provides essential support for applying high-precision data to refine soil erosion modelling.

Keywords: soil erosion modelling; parallel computing; Hadoop; MapReduce; GIS

---

1. Introduction

Soil erosion models are a useful tool for predicting the amount of soil erosion, guiding the allocation of soil and water conservation measures, and optimizing the utilization of water and soil resources in basins. As soil erosion has become a global environmental problem influenced by both natural and human factors, soil erosion modelling is a research topic that has drawn widespread attention around the world. Since the last century, researchers have established various soil erosion models from the prototype experiments and observations. In the early stages, the empirical statistical models were developed based on statistical analyses of observational data collected in plot and small watersheds with sheet and rill erosion. Examples of empirical models include the universal soil loss equation (USLE) [1] or its derivatives (e.g., the revised universal soil loss equation, RUSLE) [2] and Chinese soil loss equation (CSLE) [3,4]. These models have often been applied for on-site soil erosion estimates without considering the spatial heterogeneity of the soil erosion process [5]. With advances in geographic information system and remote-sensing technology, models and predictions of soil erosion have been developed from empirical statistical models. Now, many models are spatially distributed or physically based models. The spatially distributed models account for watershed heterogeneity, as reflected by the land use, soil types, topography, and rainfall, measured in the field or estimated through digital elevation models (DEMs) or remote-sensing images, and estimates of the sediment yield are generated in various spatial domains [6]. The Water Erosion Prediction Project (WEPP) model is a well-known and commonly used example; it is a process-based, spatially distributed parameter, and continuous simulation and erosion prediction model [7]. From a computational perspective, soil erosion models, especially spatially distributed models, all involve raster data, such as precipitation, topographic, soil, and vegetation data. Moreover, such models include various computational subprocesses and are characterized by large data volumes and intensive computational tasks. These computations, which involve exceptionally detailed high-resolution soil erosion modelling, are limited by the data storage and computing speed of the corresponding computing platforms.

In recent years, with the rapid development of information technology, massive amounts of data have been produced in different fields. To analyse and utilize these big data resources, related parallel computing methods must be used. Conventional methods usually include high-performance computing platforms (HPCs) and parallel programming models, such as message-passing-interface (MPI) and open multi-processing (OpenMP), to split the computational tasks and combine the final computing results [8]. Additionally, with the development of graphics processing hardware, the ability of graphics processing units (GPUs) has been improved, leading to the development of the computing unified device architecture (CUDA) and open computing language (OpenCL) parallel computing framework [9]. In the GIS and Remote Sensing (RS) field, Guan et al. [10] developed the parallel raster processing library (pRPL) for raster analysis in GIS. Integrated data management class, a Geospatial Data Abstraction Library (GDAL)-based raster data Input/Output (I/O) mechanism and a static and dynamic load-balancing mode were added to pRPL 2.0. Miao et al. [11] implemented a parallel GeoTIFF I/O library (pGTIOL) based on an asynchronized I/O mechanism that displayed better performance than using GDAL as the I/O interface. Both the pRPL and pGTIOL are MPI-enabled libraries. Qin et al. [12] designed and implemented a set of parallel raster-based geocomputation operators (PaRGO) to overcome the poor transferability of parallel programs. The PaRGO is compatible with three types of parallel computing platforms: GPUs, the MPI, and OpenMP. This approach makes the details of the parallel programming and the parallel hardware architecture transparent to users. Zhang et al. [13] introduced a parallel approach to quadtree construction and implemented on general purpose GPUs. A performance test yielded a significant speedup for the studied tasks. Although these libraries rely on parallel computing capabilities for raster data processes, they are limited by particular hardware environments (GPU/multi central processing unit (CPU)) and software frameworks, making them difficult to scale or expand.

The MapReduce parallel computing model for big data analysis developed by Google performs the parallel processing of massive amounts of data through cheaper server clusters [14]. This model has advantages such as few hardware requirements, rapid scaling, and easy modelling [15]. After the launch of the Hadoop open-source platform, MapReduce has been extensively used in big data processing and has formed a complete ecosystem [16]. There are several platforms for spatial big data processing and analysis such as Hadoop-GIS [17], SpatialHadoop [18], ST-Hadoop [19], GeoSpark [20], MrGeo [21], GIS Tools for Hadoop [22], Geotrellis [23] and others. In specific areas of spatial data analysis, a series of Hadoop application cases has also emerged. In the climate data analysis area, Li et al. [24] proposed a spatiotemporal indexing method that can effectively manage and process large climate datasets using the MapReduce programming model in the Hadoop platform and built a high-performance query analytical framework for climate data using the Structured Query Language (SQL) style query (HiveQL) [25]. Gao et al. [26] harvest crowd-sourced gazetteer entries running on a spatially enabled Hadoop cluster. Similar research areas include atmospheric analysis [27], large-scale Light Detection and Ranging (LiDAR) data analysis [28], remote sensing [29], trip recommendation [30], and others [31,32]. These platforms and cases are suitable for different spatial data and can be applied in different applications and domains. In this paper, we focus on integrating the Hadoop platform with GIS for parallel computing involving soil erosion modelling.

With the rapid development of geographic information technology, three-dimensional laser scanning technology can be applied to obtain high-precision geographic information data such as DEM data at the basin scale. As the resolution of raster data has increased, the data volume has exponentially grown, thus demanding a higher requirement for refined large- to medium-scale soil erosion modelling in basins. The Hadoop parallel computing platform can be used to build a parallel computing framework for soil erosion modelling and then be tightly integrated or coupled with GIS for refined soil erosion modelling. In this study, we implemented a parallel computing framework (Mr4Soil) integrated with the GIS platform. The framework includes three layers: the methodology layer, algorithms layer, and application layer. We designed two types of parallel computing strategies as rows-oriented and sub-basin-oriented methods using a spatially distributed model for annual sediment yield associated with soil erosion. We developed parallel algorithms for soil erosion modelling in the Hadoop platform based on the MapReduce parallel computing method. A series of experiments showed that Mr4Soil yielded better performance than other conventional serial programming methods. Thus, the proposed approach provides a solution for the refined soil erosion modelling on the Hadoop platform, which is integrated with a GIS platform.

2. Mr4Soil Framework Overview

2.1. Model Description

This study uses the annual sediment yield model for the large- to medium-scale basins developed by the Yellow River Institute of Hydraulic Research. Adopting Chinese Soil Erosion Equations and implementing a spatially distributed method, the model is based on the spatial discretization of the basin and comprehensively considers influences such as precipitation, soil, and topography. Therefore, the model is suitable for estimating the annual sediment yield of large- to medium-scale basins. The model formula [33] is as follows:

$$A_i = R \times K \times LS \times B \times E \times T \times g \tag{1}$$

where A is the annual soil erosion in raster cell i; R is the rainfall erosivity, which is calculated at each rain gauge station in the basin by using long-term mean monthly rainfall data; K is the soil erodibility factor, and E is the soil and water conservation practice factor. The values of K and E are related to different soil types and the distribution of soil and water conservation engineering measures in the river basin. Among them, the K factor mainly considers the composition of soil organic carbon and the particle sizes of different soil types, and E is estimated using statistical data in different administrative regions in the river basin. LS is a topographic factor that can be calculated using a CSLE model based on the extracted slopes and slope length parameters. B is a vegetation factor that reflects the influence of surface vegetation on soil erosion. B can be estimated for different land uses and vegetation coverages. T is the tillage factor, which reflects the relationship between soil erosion and farming methods, and it can be determined according to the reduction in soil erosion as a result of contour ploughing under different slope conditions. Additionally, g is the gully erosion factor, which reflects the degree of erosion by surface runoff in the river basin, and the annual mean value can be calculated by considering the slope conditions. The calculation methods for each factor are shown in Tables 1 and 2.

Table 1. Soil erosion model factors calculation method.

Factors	Calculation Method	Parameters Notes
R	$R = 0.183 F_F^{1.996}, F_F = \frac{1}{N}\sum\limits_{i=1}^{N}(\sum\limits_{j=1}^{12} P_{i,j}^2)/(\sum\limits_{j=1}^{12} p_{i,j})$	$P_{i,j}$ is the rainfall of i year, j months, N is the number of years
K	$K = \left\{0.2 + 0.3\exp\left[0.0256 SAN\left(1 - \frac{SIL}{100}\right)\right]\right\} \times \left(\frac{SIL}{CLA+SIL}\right)^{0.3} \times \left[1.0 - \frac{0.25C}{C+\exp(3.72-2.95C)}\right] \times \left[1.0 - \frac{0.7SN1}{SN1+\exp(-5.51+22.9SN1)}\right]$	$SAN/SIL/CLA/C$ are the corresponding soil types of sand grains, powders, sticky grains and organic carbon content, $SN1 = 1 - SAN/100$
LS	$S = \begin{cases} 10.8sin\theta + 0.03, & \theta < 5° \\ 16.8sin\theta - 0.50, & 5° \le \theta < 10° \\ 21.9sin\theta - 0.96, & \theta \ge 10° \end{cases}$	θ is the slope of the raster cell
	$L = (\lambda/22.1)^m; m = \begin{cases} 0.2, & \theta \le 1° \\ 0.3, & 1° < \theta \le 3° \\ 0.4, & 3° < \theta \le 5° \\ 0.5, & \theta > 5° \end{cases}$	λ is the slope length
B	$f = \frac{NDVI - NDVI_{\min}}{NDVI - NDVI_{\max}}$	f is vegetation cover ratio, $NDVI$ is normalized vegetation index, B value is shown in Table 2
E	$E = \left(1 - \frac{\alpha S_t}{S}\right) \times \left(1 - \frac{\beta S_d}{S}\right) \times \left(1 - \frac{\omega N_{d1} + \epsilon N_{d2}}{A \times S}\right)$	$S/S_d/S$ are the area of terraced fields, silt storage dams and total land, α/β are the sand reduction coefficient of terraced fields and silt storage dams, N_{d1}/N_{d2} and ω/ϵ are the number and sand reduction quota of sand or check dam, A is the annual erosion modulus
T	$T = \begin{cases} 1.000, & \theta = 0° \\ 0.100, & 0° < \theta \le 5° \\ 0.221, & 5° < \theta \le 10° \\ 0.305, & 10° < \theta \le 15° \end{cases}, T = \begin{cases} 0.575, & 15° < \theta \le 20° \\ 0.705, & 20° < \theta \le 25° \\ 1.000, & \theta \le 25° \end{cases}$	θ is the slope of raster cell
g	$G = 1 + 1.60\sin(\theta - 15)$	θ is the slope of raster cell

Table 2. *B* factor values for different land-use and vegetation cover ratio.

Landuse	Vegetation Coverage (%)	B Factor Value	
Forest/grass	0~20	0.10 (forest)	0.45 (grass)
	20~40	0.08 (forest)	0.24 (grass)
	40~60	0.06 (forest)	0.15 (grass)
	60~80	0.02 (forest)	0.09 (grass)
	80~100	0.004 (forest)	0.043 (grass)
Construction	-	0.9	
Water	-	1	
Arable	-	0.23	

From a computational perspective, this model involves a large amount of raster analysis computations. According to the range of raster neighbourhoods involved in the analysis, the computing tasks can be classified into three types: (1) local operations, which use the calculation results of the current raster cell are independent of other cells and only related to the information about itself or raster cell at the same position in other layers; (2) focal operations, which use the calculation results of the current raster cell are related to neighbouring cells within a certain range, in which a typical neighbourhood is a 3 × 3; and (3) zonal operations, which use the calculation results of a single cell that affect those of other cells, and the affected range is an irregular zone (e.g., a sub-basin). The aforementioned tasks affect data splitting strategies for model parallel computing.

2.2. Framework Overview

According to the structure and computational requirements of the soil erosion model, the Mr4Soil parallel computing framework is designed, as shown in Figure 1. The framework includes three layers such as the methodology layer, algorithm layer, and application layer. In the methodology layer, the data splitting strategies for splitting by row and splitting by sub-basin are designed according to the three types of operational tasks first. Then we abstract six parallel computing operators in the algorithm layer. The operators consist of (1) local operation parallel operators, including the spatial interpolation, vector rasterization, and map algebra operators, which are suitable for the strategy of non-overlapping splitting by row and are used to calculate R, K, E, B, and soil erosion; (2) focal operation parallel operators, including the slope analysis and flow direction analysis operators, which are suitable for the strategy of overlapping splitting by row and are used to calculate T, g, and LS; and (3) zonal operation parallel operators, mainly involving the slope length extraction operator, which is suitable for the sub-basin splitting strategy and is used to calculate the slope length of LS. The six parallel computing operators are implemented using the Hadoop platform MapReduce library. These operators can be called during soil erosion modelling. From six parallel computing operators, functions of the model parameters and the model computations are developed and compiled as Java jar file to execute on the Hadoop platform. The methodology and algorithm layer are in the Hadoop platform. To integrate the soil modelling parallel functions with GIS platform, tools for cluster connection, data pre-processing, and model computation are implemented using the ArcPy library, Secure Shell (SSH) remote access protocol, and Python language. The remote execution of parallel computing functions in the soil erosion model is achieved in the ArcGIS environment of the local machine.

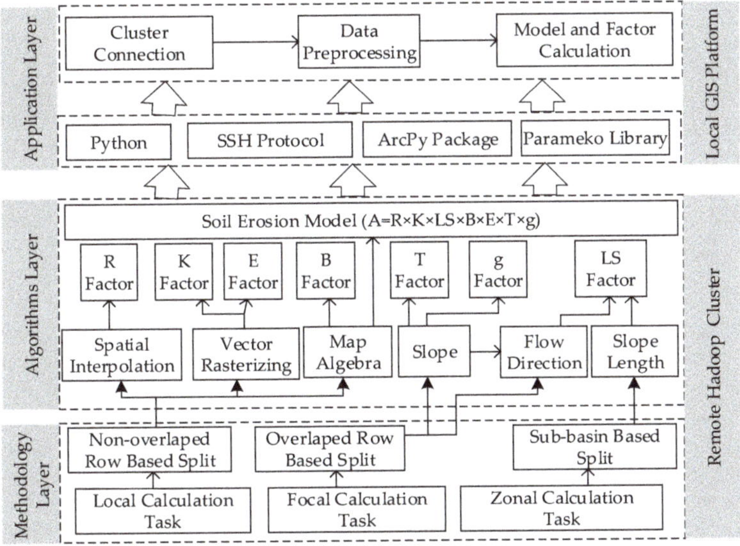

Figure 1. Overview of parallel computing tools for the soil erosion model.

2.3. Data-Splitting Methods

The Hadoop platform uses the MapReduce programming model to perform distributed computing on big data. This model provides a complete set of programming interfaces for developing distributed application programs. By splitting the input data, computations are performed for each subslice by the Mapper process, and the computational results for the subslices are then combined in a reducing process [34]. Ideally, the processing among various computing nodes after data splitting should be independent of each other to minimize the cost of communication among nodes. According to this requirement and the calculation task characteristics of the soil erosion model, we designed two data splitting strategies as follows.

(1) Split by row. In this splitting strategy, according to the number of computational nodes in parallel clusters, the raster data are split according to the number of rows. The raster data in the

corresponding row are assigned to the corresponding computing node. The number of raster rows assigned to each node is calculated using Equation (2).

$$h = \begin{cases} RC/N & if\ mod(RC, N) = 0 \\ int(RC/N) + 1 & if\ mod(RC, N) \neq 0 \end{cases} \tag{2}$$

where h is the number of raster data rows for each subslice, RC is the total number of rows of raster data, N is the number of computational nodes, and the *mod* is a residual function. For local operations, the calculation results for a raster cell are not related to and do not affect the results of other cells. Therefore, there is no need to consider the data in adjacent cells after the split, meaning that there is no overlap in splitting (Figure 2). For example, the map algebra operation can be finished with the non-overlap data splitting method because each raster cell is calculated independently of its neighbourhood raster. After the raster data input, the raster is divided into non-overlap segmentation according to the row and assigned to the calculation node in the cluster. Then each node can perform the corresponding algebraic operations on the allocated data rows at the same time. Figure 2 is an example of the multiplication of each raster cell.

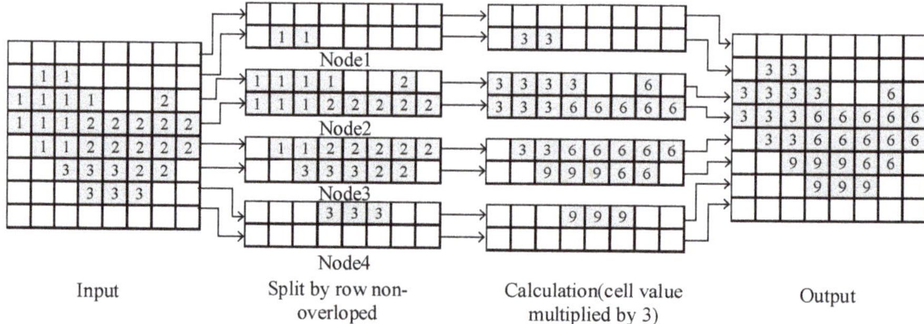

Figure 2. Data non-overlap splitting by row with 4 calculation nodes.

For neighbourhood operations, the neighbourhood ranges are different among computations. To ensure that all cells in the corresponding neighbourhood are read at the same computational node, h must be expanded on the basis of non-overlapping splitting. Assuming that the width of a neighbourhood is W, the upper and lower boundaries of the neighbourhood are each $(W-1)/2$ rows away from the centre, meaning that there is overlapping splitting. Figure 3 is an example of summing the range of the 3×3 neighbourhood of each raster cell. With an overlap segmentation strategy, the rows of raster data allocated to each compute node are overlapped, thereby completing independently at each node. The cell in the 3×3 window is calculated, and finally, the results of each node are combined and output.

(2) Split by sub-basin. In the calculation of some parameters in the soil erosion model (such as the slope length factor), the calculation results of the corresponding raster cell are related to the values of other cells within a certain range. In contrast to neighbourhood operations, this range is an irregular area (such as a sub-basin). The parallel computation of these factors is performed using the sub-basin splitting strategy. That is, using DEM data in the basin, sub-basins are divided according to watershed lines; then, the sub-basins are assigned to the corresponding nodes according to their spatial adjacent relations (Figure 4). When the basin is divided into sub-basins, the sub-basins can be properly merged or subdivided under the correct flow accumulation in a watershed. To avoid large differences in the data volume of nodes, which influences the overall computational efficiency, the sub-basin area assigned to each node should be roughly equal so that the computing load of each node is balanced. As shown in Figure 4, the watershed is divided into 67 sub-basins. Since the cluster has four computing nodes, these sub-basins are merged according to the adjacent relationship to obtain four

polygon ranges. The point in polygon spatial relationship is identified for each raster cell and get the polygon ID value. Then the cells are assigned to different computing nodes according to corresponding polygon ID to complete parallel computing tasks. Using the data input class *Nlineinputformat* and *Keyvalueiputformat* class in the Hadoop platform MapReduce library [35], the input raster data could be partitioned by configuring the raster rows, worker nodes, and the sub-basin boundary parameters.

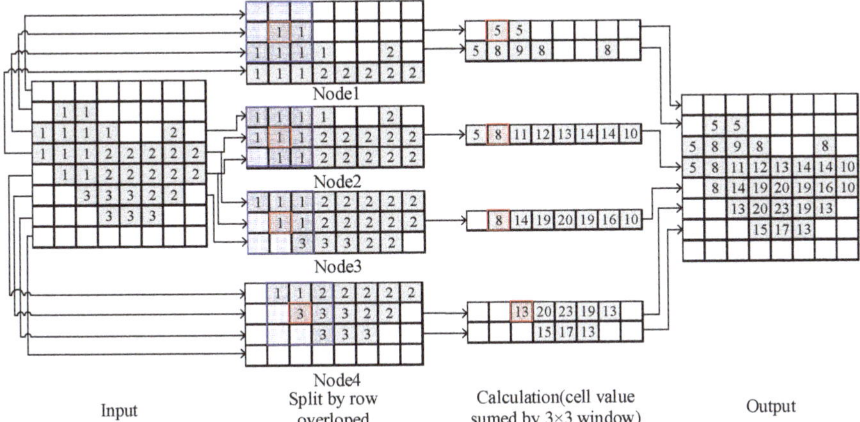

Figure 3. Data overlap splitting by row with 4 calculation nodes.

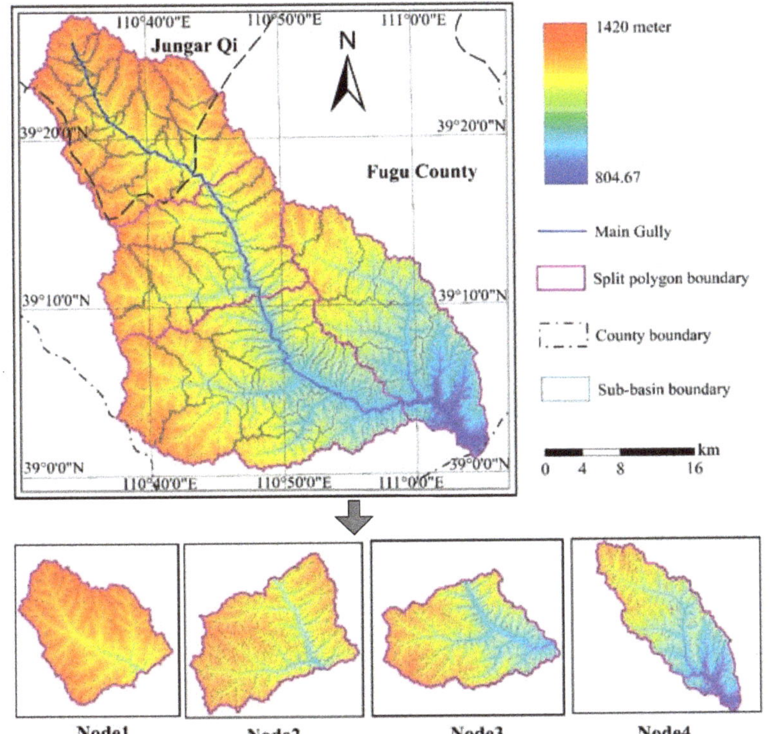

Figure 4. Data splitting by sub-basin with 4 calculation nodes.

3. Parallel Algorithms Design

The parallel algorithms of model consist of 3 levels: (1) the parallel model operators, which are used to define the basic analysis algorithm; (2) the parallel factor algorithms, which are used to achieve parallel computing of different model factors; and (3) the parallel model algorithm, which is used to calculate soil erosion according to Equation (1). The parallel model operators are at the bottom level. The model factor and model calculations can be completed by calling the corresponding operators. Based on the data structure definition and data preprocessing, this section focuses on the algorithm design for parallel model operators.

3.1. Data Structure and Data Preprocessing

The Hadoop platform supports the reading and writing of structured text data and binary data. To simplify the calculations, the (key, value) pair in text format is applied to define the data structure involved in parallel model computing. There are two types of input data used in model computations: vector and raster data. Vector data mainly include rainfall data collected at rain gauge stations (point) in the basin, soil type data, soil and water conservation measure data (polygon), and sub-basin boundary data (polygon). The volume of these data is relatively small, and there is no need for splitting in the calculation. Vector data are converted into a text format supported by the Hadoop platform. In this study, vector data are converted into GeoCSV format. The geographic entity code is used as the key. Spatial data, which are stored in well-known text (WKT) format, together with property data are used as the value and can be directly read by the Hadoop distributed file system (HDFS) API and used in calculations. The raster data mainly include the basin DEM, land-use data, and normalized difference vegetation index (NDVI) data. First, these data are converted into two-dimensional text data. Then, two types of (key, value) pairs are defined: (1) an entire row is used as a (key, value) pair, where key is the row number and value is the set of the corresponding raster cell values of the row; and (2) a raster cell is used as a (key, value) pair, where the row number and column number of a raster cell are defined as the composite key, and the value of the raster cell is defined as the value. The raster data are preprocessed, and the corresponding (key, value) pairs in text format are generated and used as input data in different parallel algorithms.

3.2. Parallel Algorithms Based on the Row-Splitting Method

The row-splitting method is suitable for local and neighbourhood operations. By using an entire row as a (key, value) pair for input data, the calculation process includes 2 steps: (1) in the map step, the splitting function is defined according to the key of the row of the input raster data, and the raster data row is assigned to the corresponding computing nodes; (2) in the reduce step, based on the different analysis algorithms, calculations are performed involving the slices of data rows. The results of the slices are sorted according to the key, combined and output. In local operation tasks, there is no need to consider the overlap among slices. The (key, value) pair raster data in text format can be directly used in calculations. In focal operation tasks, the corresponding raster rows must be duplicated in the first and last rows of each slice according to the size of the neighborhood windows. Herein, the spatial interpolation parallel operators based on the inverse distance weighting (IDW) method are used as an example. The input parameters of the parallel algorithm include: (1) the observation station data as GeoCSV, which stores the station position and its observation value. The data volume is small and can be used for each computational node without data splitting. (2) The row key-value pair raster data in text format (the initial value of a raster cell is set to -1). The key is row number, and the value is the cell list of the row. Spatial interpolation is the process of calculation each raster cell value based on the observation station data. The calculation steps are as follows (Figure 5).

(1) Data input. Read the key-value text of the row, and parse the row number and the raster cell list; read and parse the GeoCSV text to get the station position and value.

(2) Data splitting. The number of raster rows allocated by each node is calculated according to formula (2) and cluster size, and then the rows are assigned to the corresponding computational node by customizing the MapReduce splitter (partitioner) to perform non-overlapping row segmentation according to the line number.

(3) Data parsing and distance calculation. The coordinates of the raster cell center are calculated from the cell width and row/column number. The distance from the raster cell and all of the station are measured to generate a set $\{(d_i, v_i)\}$, where d_i is the distance and v_i is the observation value.

(4) Spatial interpolation. According to the IDW interpolation method, the interpolation points within a certain range of the current raster cell by the distance are selected. The interpolation calculation is weighted by the distance, and the values of the corresponding raster cells (c_{ij}) are updated.

(5) Output. A list is generated based on the interpolation result of each raster cell of the row, and written to the HDFS along with the row number (Figure 6).

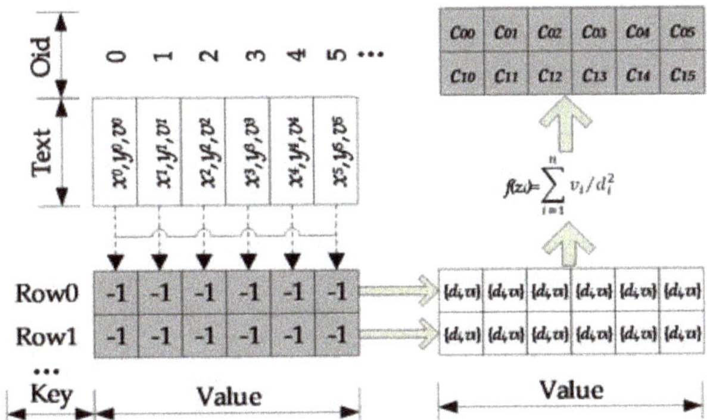

Figure 5. Parallel algorithm example for spatial interpolation.

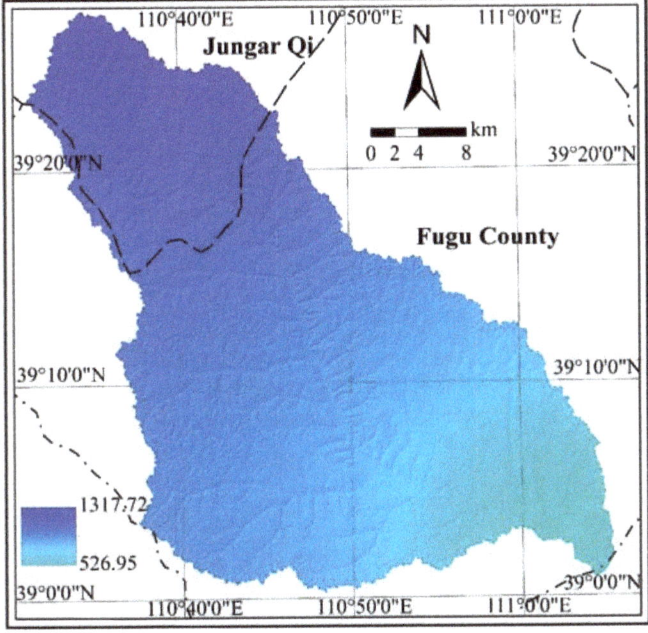

Figure 6. R Factor using row splitting based parallel algorithm.

3.3. Parallel Algorithms Based on the Sub-Basin Splitting Method

Among the topographic factors of the model, slope length is a key parameter that affects soil erosion. The serial computing process of the slope length includes steps such as slope and flow direction determination, flow accumulation, cut-off detection, cell slope length determination, and cumulative slope length calculations. Among these steps, the flow accumulation and cumulative slope length calculation steps involve cumulative process and different raster cells according to the flow path. Parallel computations cannot be directly performed following the row-splitting strategy. The flow paths include two subsets, namely, the flow paths in each sub-basin and the main gully in the basin. Flow accumulation can be calculated independently in the sub-basins, whereas flow accumulation in the main gully of the basin involves fewer raster cells, and thus, no data splitting is needed. Flow accumulation in the main gully can be completed directly in the reduce phase after the sub-basin calculations are performed. Based on the analysis above, the sub-basin splitting strategy can be used to parallelize slope length extraction. The algorithm can be implemented by using the master-slave parallel computing processes in MapReduce. Slave computing processes mainly refer to subprocesses in which only some of the data are involved and no global data need to be considered. The computing results serve as intermediate data that are input into the master computing process. The input data of the algorithm are the raster-based (key, value) pairs in text format (DEM, slope and flow direction) and the GeoCSV data of the splitting polygon which store the polygon ID and boundary coordinates. The main steps are as Figure 7 shown.

(1) Data input. Read the key-value text of the raster, parse the row and column number of the cell and calculate the cell center coordinates. Read and parse the GeoCSV text to get the splitting polygon ID and boundary coordinates.

(2) Splitting polygon identification and data splitting. The ray casting method for point-in-polygon testing is used to determine the splitting polygon in which the cell is located, and get the polygon ID to customize the partitioner to perform sub-basin splitting. Then the raster cells are sent to different computational nodes.

(3) Slope length calculation for cells in the sub-basin. For each raster cell assigned to the worker node, the flow accumulation, cut-off detection, cell slope length, and cumulative slope length are calculated using the serial calculation method of the slope length; and output the cumulative slope length of each cell.

(4) Slope length calculation for cells in the main gully. According to the result of the flow accumulation in (3), the main gully cell is determined. The flow accumulation, cut-off detection, cell slope length, and cumulative slope length of these cells is computed by the same process as step (3).

(5) Combination and output. The slope length for cells in the sub-basin and the main gully is merged with the rules as follows: if the cell is in the main gully, the slope length generated by step (4) are the slope length value of the cell; otherwise is the step (3). The slope length of the complete watershed is written to HDFS. The LS factor is then calculated according to the corresponding method in Table 1 (Figure 8).

According to the two parallel algorithms described above, spatial interpolation, vector rasterization, map algebra, slope gradient, flow direction, and slope length operations are implemented using the MapReduce programming interface. Based on the model factors and model structure, the corresponding parallel operators are called to calculate model factors and perform model calculations in the basin.

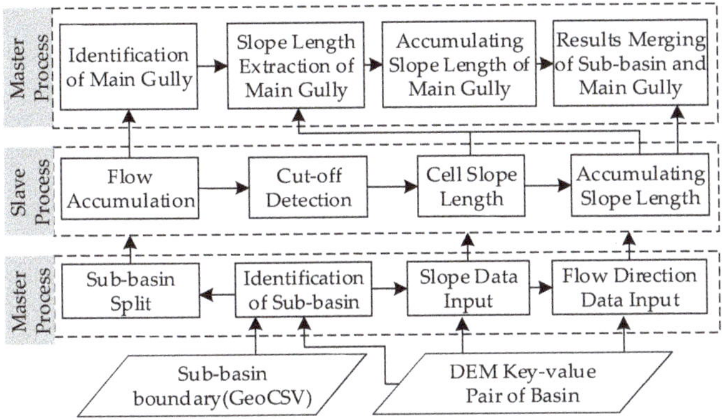

Figure 7. Diagram of parallel algorithm for slope length.

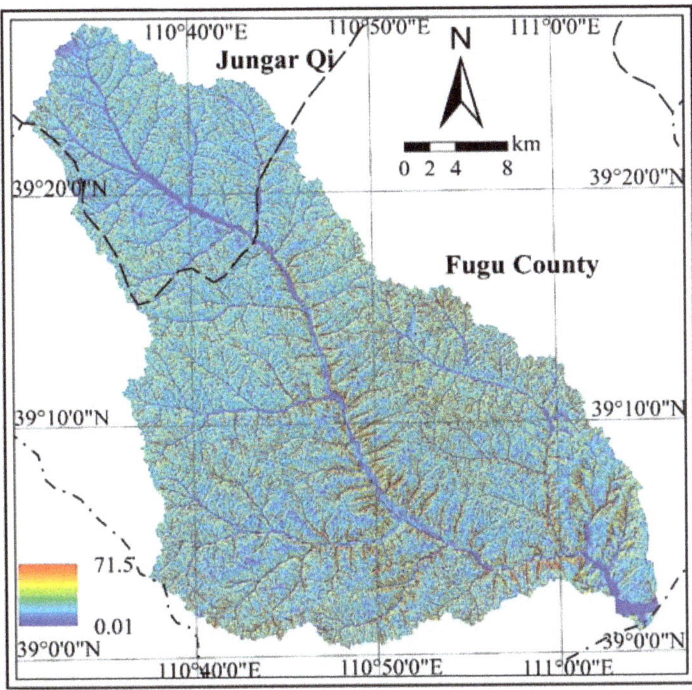

Figure 8. *LS* factor using sub-basin splitting based parallel algorithm.

4. Toolbox Development Integrated with Geographic Information System (GIS)

The Hadoop platform is operated on a Linux parallel computing cluster that consists of multiple computing nodes. Thus, multiple Linux and Hadoop commands must be executed to complete the corresponding computing tasks. The process also involves uploading or downloading model data to/from clusters, which complicates the parallel computing tasks in the model described above. To improve the usability and ease of operation of parallel computing functions in the model, a geoprocessing toolbox is compiled from the model parallel computing functions using a Python script based on the ArcGIS platform. This toolbox allows for the direct calling and the visualization of the computational results in the ArcGIS platform, thus achieving the integration of the parallel computing functions of the model with ArcGIS.

As the toolbox runs on the ArcGIS platform of the local machine, the SSH (secure shell) protocol is used to complete the communication between the local machine and the remote cluster. The SSH protocol is a method of securing remote login operations from one computer to another. The Linux platform provides an SSH access protocol for remote login and other operations. On the Windows platform of the local machine, we use Paramiko, an SSH access library in Python, to interact with the Linux platform. Paramiko provides two types of objects: SSHClient, which is used to implement login to remote hosts and perform various command operations, and SFTPClient, which is used to implement upload and download files. The calculation pipeline of the Mr4Soil is shown in Figure 9. First, according to the internet protocol (IP) address of the remote host, network port, and other information, a user can connect to the master node of the Hadoop cluster via the SSHClient and SFTPClient objects on the local machine and then convert the data in GIS format to GeoCSV format or a key-value text file through the ArcPy package. Second, the converted data can be uploaded to the master node by calling the put method of SFTPClient. Next, these files on the master node are transferred to the HDFS of the Cluster using Hadoop platform file operation command and the EXEC_COMMAND function of SSHClient. Then, the corresponding MapReduce program is executed using the Hadoop platform Java Archive (JAR) operation command in the same way to compute the various factors and perform soil erosion calculations. Finally, the user can download the results to the local machine and convert them to GIS format and use the ArcGIS platform for browsing and viewing.

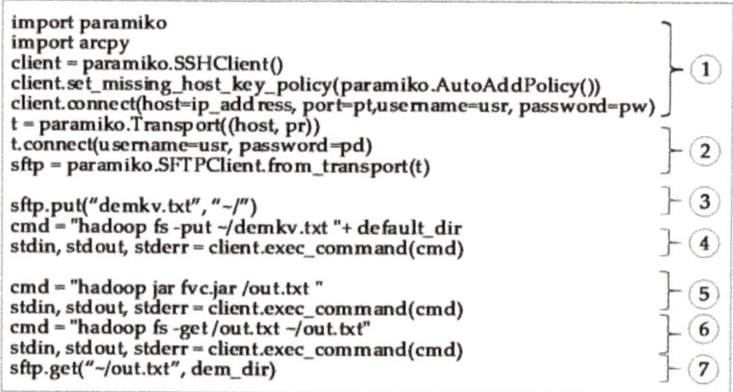

Figure 9. The calculation pipeline of the framework: connection with master node by SSHClient and SFTPClient in and ; uploading the input data to master node in and Hadoop HDFS in ; the model calculation in , and the output data download to master node in and local machine in from HDFS.

There are three tools in the application layer of the framework (Figure 10a). (1) The cluster connection tool. Using the Paramiko module and the SSH protocol to login to the cluster remotely, files can be uploaded and downloaded through SFTP (SSH file transfer protocol). For a flexible connection method, the cluster login IP and network service port are set as login parameters. Parameter passing occurs through the ArcGIS customization tool. Remote login to the clusters in the Windows environment is then achieved (Figure 10b). (2) The data preprocessing tool. This tool is mainly used to achieve two-way interoperability between native GIS data and text data for the HDFS of the cluster. The tool converts GIS vector data to GeoCSV format and GIS raster data to key-value pair text format using the data access and conversion functions in ArcPy. Then, the data processing tool uploads the data to HDFS to read in model operations through the SFTPClient object of the Paramiko library (Figure 10c). (3) The model computing tool. Model computation is a process based on the calculation function of each model factor and soil erosion. This process compiles these functions into JAR packages, uploads the packages to the cluster master nodes, and calls the packages. The model computing tool encapsulates this process. After the user specifies the computing task, the corresponding JAR package is executed by the SSHClient object, thereby completing the model computation and writing the results to the HDFS (Figure 10d).

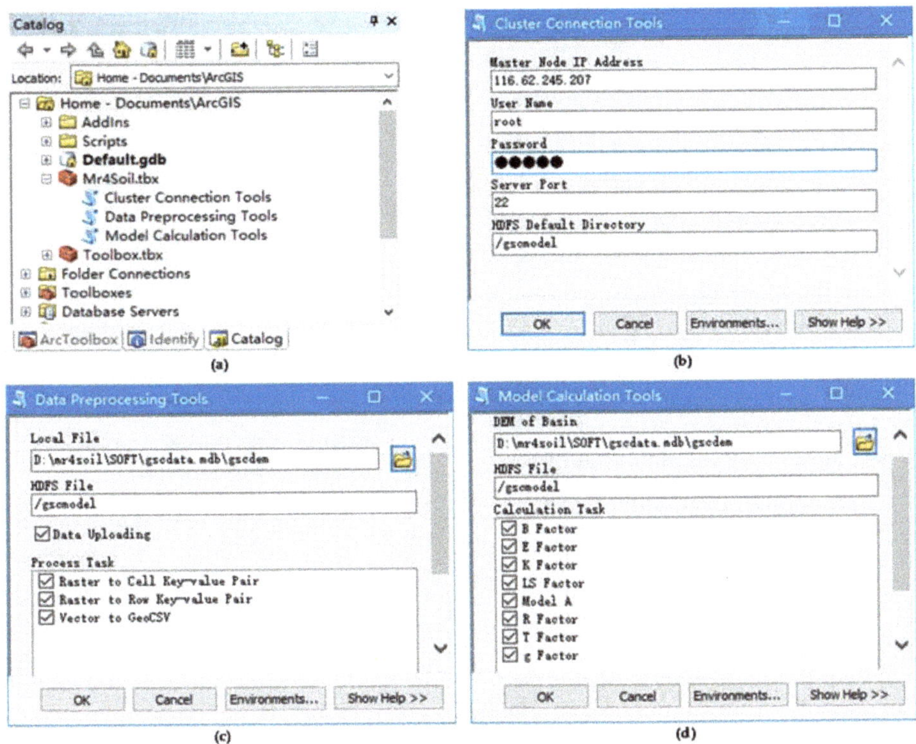

Figure 10. Mr4Soil toolbox and three tools.

5. Experiment and Test

5.1. Data Sources and Experimental Environment Configuration

The Loess Plateau is an area of severe soil erosion in China. This paper selects the Gushanchuan basin on the Loess Plateau as a typical research area (Figure 11) for Mr4Soil application and testing. Gushanchuan is a primary tributary on the right bank of the middle reaches of the Yellow River. It originates in Jungar Qi, Inner Mongolia, and flows through Jungar Qi and Fugu County, Shaanxi Province, and into the Yellow River in Fugu Town. The length of the main channel is 79.4 km, and the average slope ratio is 5.4‰. The Gushanchuan basin (E110°32′24″~E111°05′24″, N39°00′00″~N39°27′36″) is located on the southeast slope of the Ordos Plateau, belonging to the first portion of the loess hilly and gully region. The total area of the basin is 1272 km², of which 254 km² is in Jungar Qi, and 1018 km² is in Fugu County. The basin is located in the transition zone between Mu Us desert and loess hilly and gully region, and the geomorphology is typical of the loess hilly and gully region. A small portion of the upstream region is covered by a loess sand area, and the gully density is 2.91 km per sq. km. The soil types in the basin are mainly loessal soils, which account for approximately 66.67% of all soils, followed by chestnut soil, which account for about 26.74% of all soils. The watershed is located in the semi-arid continental monsoon climate zone, with an annual average temperature of 7.3 °C and an average annual rainfall of approximately 410 mm.

Precipitation varies significantly each year and is unevenly distributed throughout the year, mostly occurring in the form of torrential rains. Rainfall during the flood season (June-September) can account for 80% of the annual rainfall. The main vegetation types in the basin are *Stipa bungeana* and *Artemisia*. The type of soil erosion in the watershed is mainly water erosion, and the average annual erosion rate is 16,800 tons per sq. km, and the annual sediment transport volume is 21.39 million tons.

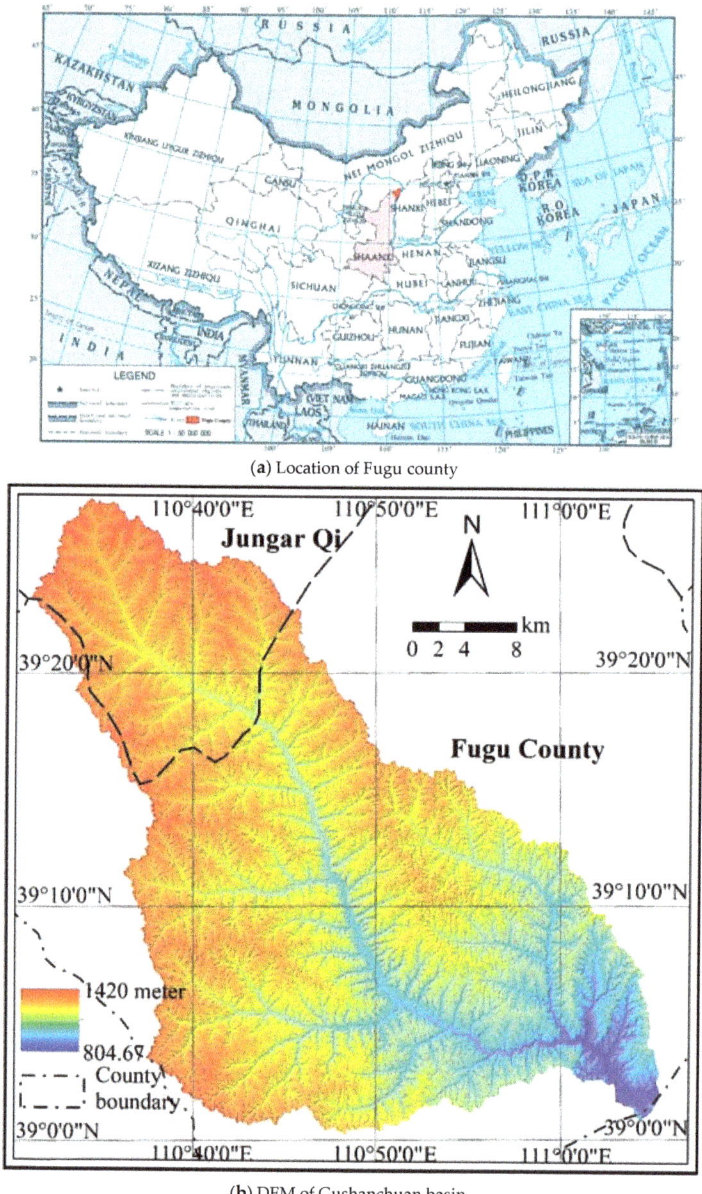

(**a**) Location of Fugu county

(**b**) DEM of Gushanchuan basin

Figure 11. Location and digital elevation model (DEM) of Gushanchuan basin.

We calculated the model factors and performed model calculations for the Gushanchuan basin using the Mr4Soil framework. To test the computational efficiency, four spatial resolutions of 30 m, 10 m, 5 m, 2.5 m were selected to calculate the seven model factors and soil erosion quantity. The model input data involved vector data and raster data. The vector data includes: (1) the soil types in the basin and the soil and water conservation engineering measures (polygon), which were used to calculate the K and E factors by the vector to raster operator. (2) The R factor was calculated using the site-specific basin rainfall data (point) by the spatial interpolation operator. (3) The sub-basin boundary data (polygon) were used for the sub-basin splitting method when the slope length was calculated. The vector dataset is small, which were converted directly to GeoCSV format without splitting. The raster data include the DEM, land-use type and NDVI data for the basin, and these data are used to calculate LS, B, T, g factors. The data volume of these raster data differs at various spatial resolutions, and the data should be partitioned by row-based or sub-basin-based methods before calculations using the parallel algorithms. The amount of raster data at different resolutions is shown in Table 3.

Table 3. The datasets volume and row/column count for the DEM.

Spatial Resolution	Row Count	Column Count	Data Volume	VALUED CELLS
30 m	1692	1666	10.75MB	1,564,189
10 m	5076	4998	96.78MB	14,077,704
5 m	10,152	9996	387.11MB	56,310,815
2.5 m	20,304	19,992	1511.46MB	225,243,260

The Alibaba cloud platform is used to configure the framework runtime environment. The Hadoop cluster that runs in the Alibaba cloud consists of 5/9/17 hosts, including 1 master node and 4/8/16 worker nodes. The node hardware is configured as a 4-core CPU, with 32 GB RAM and a 200 GB solid-state disk hard drive. At each node, the Ubuntu16.04 operating system is installed, the Java runtime environment is configured, and Hadoop 2.7.2 is adopted. The cluster master node has a public network IP address and supports remote login access. The cluster network environment is a Gigabit local area network. The model computing tool runs on the ArcGIS platform of a local Windows host and interacts with the Hadoop cluster based on the network access protocol. To analyse the parallel computing performance of the tools, a single workstation with the same configuration is used for serial computations. Performance analysis is conducted based on the parallel acceleration ratio, which is defined in Equation (3).

$$S_p = T_s / T_p \tag{3}$$

where S_p is the parallel acceleration ratio; T_s is the serial computing time; and T_p is the time required when parallel computing is performed using p computing nodes, which is measured by the time between calculation job submitted and finished in the Hadoop platform. Table 4 shows the information for the experimental environment.

Table 4. The test environment configuration.

Machine Role	System Configuration
Master node of cluster	Ubuntu16.04; JDK1.6; Hadoop 2.7.2; 64 GB Memory; 200 GB Hard disk
Worker node of cluster	Ubuntu16.04; JDK1.6; Hadoop 2.7.2; 32 GB Memory; 200 GB Hard disk
Local machine	Windows 7; ArcGIS 10.3; 32 GB Memory; 500 GB Hard disk

5.2. Parallel Acceleration Ratio for Data with Different Spatial Resolutions

We calculated the parallel acceleration ratio for spatial resolutions of 30 m, 10 m, 5 m, and 2.5 m, as shown in Figure 12. As data resolution increases, the parallel acceleration ratio of the algorithm significantly increases, and the computational efficiency is improved. The larger the data volume is, the more obvious the improvement and the higher the parallel acceleration ratio of the algorithm. At 30 m and 10 m resolutions, the average acceleration ratios of three clusters are 0.82/0.93/1.01 and 1.78/2.10/2.76, respectively. Additionally, at 5 m and 2.5 m resolutions, the average acceleration ratios of three clusters are 3.55/4.27/4.89 and 5.23/6.42/7.59, respectively. The maximum acceleration ratio is 12.52 (K factor) at a 2.5 m resolution. It is worth noting that in the case of the 30 m resolution, the acceleration ratios of some factors in the parallel algorithm are less than 1.0, mainly due to the communication requirements during data splitting and combining. A linear regression of the calculation time using the serial and parallel computing also confirms the strong positive relations between them. The goodness of fit (R^2) of the regression models for the three types of clusters ranges from 0.78 to 0.87. Therefore, the parallel computing method in the soil erosion modelling with higher spatial resolution data improves the computational performance of different tasks.

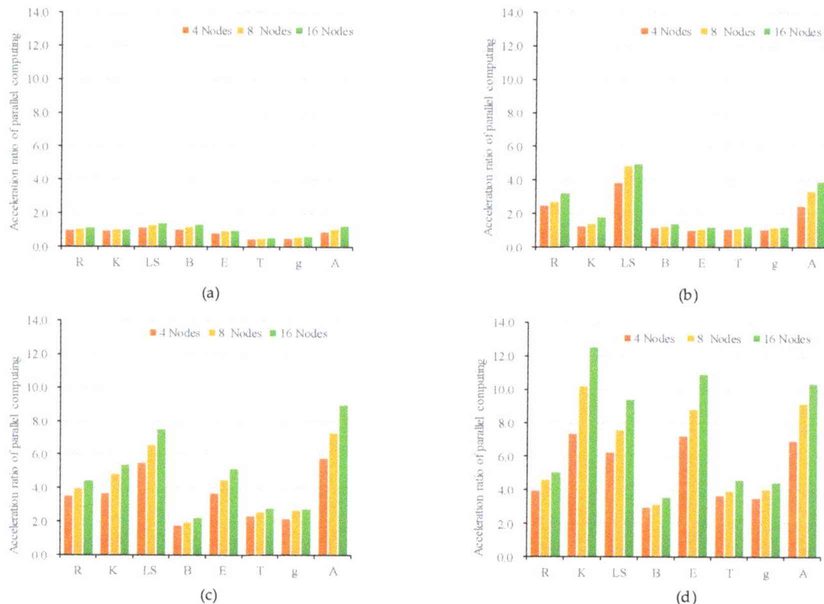

Figure 12. Parallel acceleration ratio for data with different spatial resolution. (**a**) raster cell width with 30 m; (**b**) raster cell width with 10 m; (**c**) raster cell width with 5 m; (**d**) raster cell width with 2.5 m.

5.3. Time Consumption Test for Different Calculation Tasks

Based on the different clusters with 4, 8, and 16 work nodes, we measure the time consumption of calculations for eight tasks in one model with seven factors, as shown in Figure 13. In general, the calculation times of different clusters vary for each task. At the same spatial resolution, the calculation time gradually shortens with the expansion of the cluster scale. The calculation time differences increase as the spatial resolution and data volume increase. Based on three sizes of Hadoop clusters, the average time required to complete eight computing tasks for 30 m resolution data is 0.70/0.61/0.57 (min), and the average times for the corresponding 10 m, 5 m, and 2.5 m resolution datasets are 1.41/1.23/1.09, 2.08/1.76/1.54, and 3.17/2.64/2.24, respectively. When the cluster size is held constant, the calculation time difference is relatively small because the volume of data is small. For example, when the resolution is 30 m, 10 m, and 5 m, the calculation time difference is approximately 0.45~0.72, and at a 2.5 m resolution, the difference is 0.70~1.08. For the specific calculation task, the *LS* factor has the most significant difference in calculation time of different cluster scales, and the average calculation time of *LS* at the three scale clusters is 3.49, 2.88 and 2.43

respectively. Notably, the calculation time for the 2.5 m resolution data decreases the most, and as the number of work nodes increases from 4 to 8 and 16, the calculation time decreases by 1.50 and 1.34 respectively. In addition, the average calculation time of model A significantly decreases, and the average calculation time is 2.61, 2.01 and 1.71. The time consumption for the *LS* factor is relatively higher than model A. These two types of computing tasks all involve a variety of input data, but the *LS* factor calculation has more subprocesses (such as cell slope length, flow accumulation, cut-off detection, and main gully calculation tasks) and master–slave communication. In addition, the calculation time differences among the other six computational tasks are relatively small. The parallel calculations of the Hadoop cluster provide a distinct advantage for the soil erosion modelling with large data volume and high computational complexity.

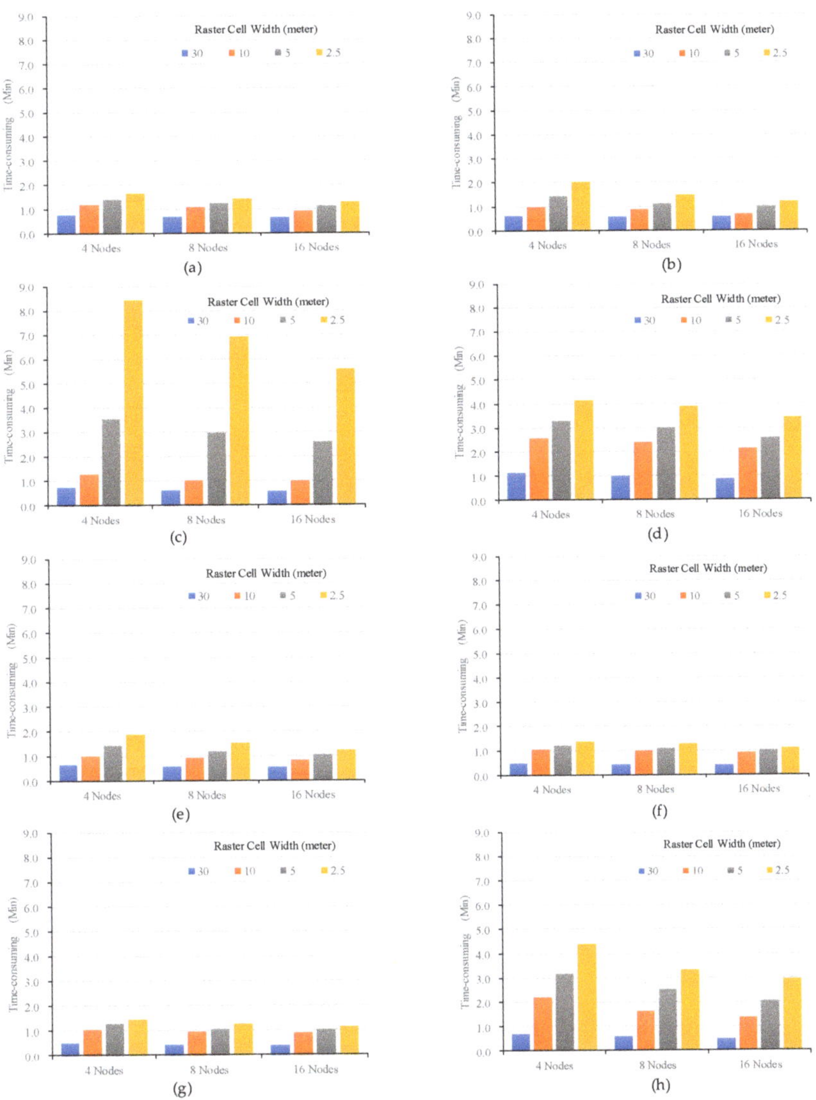

Figure 13. Time-consumption with different calculation tasks. (**a**) *R* factor; (**b**) *K* factor; (**c**) *LS* factor; (**d**) *B* factor; (**e**) *E* factor; (**f**) *T* factor; (**g**) *g* factor; (**h**) model *A*.

We take the slope calculation as an example to examine the calculation time of the GPU, CPU, and user-defined aggregations (UDA) using the same configuration parameters (Table 4) of the cluster worker nodes. The NVIDIA Telsa M40 graphic card is used to calculate the slope. The UDA test uses the PostgreSQL 11 database platform, which supports setting the number of parallel processes and using the SQL language for UDA parallel computing. The number of working processes for UDA calculation is set to 16. The UDA function is written in SQL to perform parallel slope calculation. The calculation time of different scale cluster computing time and PC-based GPU, CPU and UDA is compared (Figure 14). It can be seen that in the case of a small amount of data, the GPU and CPU are faster in calculation speed, and the calculation efficiency is better than that of the Hadoop cluster and UDA; however, as the amount of data is gradually increased, the calculation time of the GPU/CPU and the UDA is rapidly increased. The cluster computing time increases slowly; in the case of 2.5 m resolution, the UDA calculation time is the longest, and the CPU is second; and with the expansion of the cluster size, the computing time of the 16 cluster worker nodes is roughly equivalent to the GPU computing time. In the case of four resolutions, the UDA method takes the longest calculation time. Although GPU computing efficiency is better than Hadoop, the cluster does not require expensive hardware such as GPU and is suitable for soil erosion modelling on cloud computing platforms.

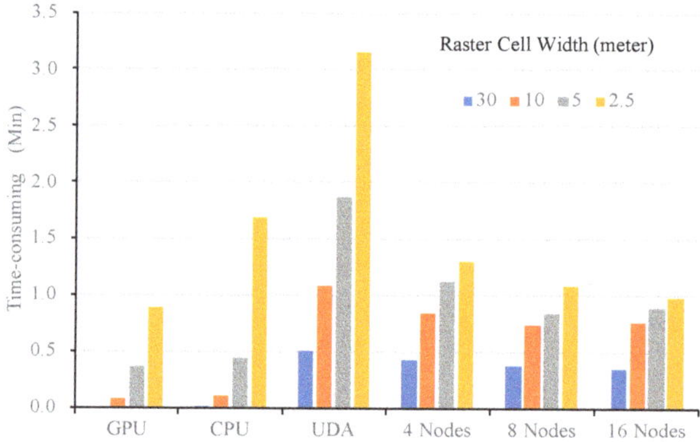

Figure 14. Time-consumption for slope calculation with graphics processing unit (GPU), central processing unit (CPU), unified device architecture (UDA), and different cluster size.

6. Summary and Discussion

With the continuous development of geographic information acquisition technology, the requirements for soil erosion modelling and calculations have increased due to the broader availability of high-resolution data. Based on the Hadoop platform, an empirical model of soil erosion for annual sediment yield is selected to implement the Mr4Soil parallel computing framework. For three types of computing tasks, including local, focal, and zonal operations, two types of data-splitting strategies (row-based and sub-basin-based) are designed. Six parallel operators are defined, including spatial interpolation, vector rasterization, map algebra, slope gradient, flow direction, and slope length operators. The corresponding parallel algorithm is designed. A geoprocessing toolbox for model calculations is developed using the Python language based on the ArcGIS platform. The performance of the framework is analysed by taking the Gushanchuan basin on the Loess Plateau, China, as an example. With increases in the data resolution and volume, the computational efficiency of different model factors significantly improves, and a higher parallel acceleration ratio is achieved. The main contributions of this paper are as follows: (1) two data-splitting methods, row-based and sub-basin-based methods, and six parallel operators for local, focal and zonal soil erosion modelling computing tasks are developed; (2) a complete parallel computing framework for soil erosion modelling based on Hadoop platform is proposed; (3) a geoprocessing toolbox that integrates the parallel computing for soil erosion modelling with the GIS platform is constructed.

After analysing the performance of the implementation of the Mr4Soil, we found that, compared to parallel computing frameworks such as MPI, GPU, and UDA, the advantages of the proposed Mr4Soil framework are as follows: first, the Mr4Soil framework does not require specialized hardware, such as GPUs, and can achieve a high computational efficiency for a large-scale basin with high-resolution data. Moreover, the framework can be directly implemented on mainstream cloud computing platforms and has high scalability and usability. Second, the Mr4Soil framework was integrated with a GIS platform using a geoprocessing toolbox. It is easy to use for GIS users and provides essential support for using high-precision data to refine soil erosion modelling. Third, the Mr4Soil framework focuses on the parallel computing of soil erosion modelling, which is fully functional for different computing tasks and model factors.

For the parallel algorithm and geoprocessing toolbox development of the integration of GIS and soil erosion modelling parallel computing, this paper only describes a situation in which the GIS platform on a desktop and spatially distributed model are used. Further studies of the following two topics should be performed: (1) the creation of a service interface for the soil erosion model with parallel computing tasks, which could be based on a web service and conveniently applied in practice; and (2) the development of a physically based soil erosion model with parallel computing based on the Hadoop platform. These findings could extend parallel modelling studies of soil erosion.

Author Contributions: Z.G. designed and developed the Mr4Soil and wrote this paper. F.Q. and L.W. are the supporters of this project and supervised this paper. C.C. and Y.L. assisted in the system development. P.F. provided key technical guidance for the system development.

Funding: This research was funded by National Key Research Priorities Program of China, grant number 2016YFC0402402; National Natural Science Foundation of China, grant number 41871316 and 41601116; Science and Technology Foundation of Henan Province, grant number 172102110402; and the Henan University Science Foundation for Young Talents.

Acknowledgments: Special thanks go to the editor and anonymous reviewers of this paper for their constructive comments.

References

1. Wischmeier, W.H.; Smith, D.D. *Predicting Rainfall Erosion Losses—A Guide to Conservation Planning*; United States Department of Agriculture: Hyattsville, MD, USA, 1978; pp. 1–58.

2. Kenneth, G.R.; George, R.F.; Weesies, G.A.; McCool, D.K.; Yoder, D.C. *Predicting Soil Erosion By Water: A Guide to Conservation Planning with the Revised Universal Soil Loss Equation (RUSLE)*; United States Department of Agriculture: Washington, DC, USA, 1997; pp. 19–38.

3. Liu, B.; Zhang, K.; Xie, Y. An empirical soil loss equation. In Proceedings of the 12th International Soil Conservation Conference, Beijing, China, 26–31 May 2002; pp. 21–25.

4. Liu, B.; Bi, X.; Fu, S. *Soil Erosion Equation in Beijing*; Science Press: Beijing, China, 2010; pp. 36–66.

5. Joris, V.; Jean, P.; Gert, V.; Anton, V.R.; Gerard, G. Spatially distributed modelling of soil erosion and sediment yield at regional scales in Spain. *Glob. Planet. Chang.* **2008**, *60*, 393–415.

6. Ramsankaran, R.; Umesh, C.K.; Sanjay, K.G.; Andreas, M.; Krishnan, M. Physically-based distributed soil erosion and sediment yield model (DREAM) for simulating individual storm events. *Hydrol. Sci. J.* **2013**, *58*, 872–891. [CrossRef]

7. Dennis, C.F.; John, E.G.; Thomas, G.F. Water Erosion Prediction Project (WEPP): Development history, model capabilities, and future enhancements. *Trans. ASABE* **2007**, *50*, 1603–1612.

8. Jin, H.; Dennis, J.; Piyush, M.; Rupak, B.; Huang, L.; Barbara, C. High performance computing using MPI and OpenMP on multi-core parallel systems. *Parallel Comput.* **2011**, *37*, 562–575. [CrossRef]

9. Craig, A.L.; Samuel, D.G.; Antonio, P.; Chein-I, C.; Bormin, H. Recent developments in high performance computing for remote sensing: A review. *IEEE J. Sel. Top. Appl. Earth Obs. Remote Sens.* **2011**, *4*, 508–527.

10. Guan, Q.; Zeng, W.; Gong, J.; Yun, S. pRPL 2.0: Improving the parallel raster processing library. *Trans. GIS* **2014**, *18*, 25–52. [CrossRef]

11. Miao, J.; Guan, Q.; Hu, S. pRPL+ pGTIOL: The marriage of a parallel processing library and a parallel I/O library for big raster data. *Environ. Model. Softw.* **2017**, *96*, 347–360. [CrossRef]

12. Qin, C.; Zhan, L.; Zhu, A.; Zhou, C. A strategy for raster-based geocomputation under different parallel computing platforms. *Int. J. Geogr. Inf. Sci.* **2014**, *28*, 2127–2144. [CrossRef]

13. Zhang, J.; You, S. High-performance quadtree constructions on large-scale geospatial rasters using GPGPU parallel primitives. *Int. J. Geogr. Inf. Sci.* **2013**, *27*, 2207–2226. [CrossRef]

14. Jeffrey, D.; Sanjay, G. MapReduce: A flexible data processing tool. *Commun. ACM* **2010**, *53*, 72–77.

15. Deepak, V. *Practical Hadoop Ecosystem: A Definitive Guide to Hadoop-Related Frameworks and Tools*; Apress: New York, NY, USA, 2016; pp. 1–32.

16. Sam, R.A. *Expert Hadoop Administration: Managing, Tuning, and Securing Spark, YARN, and HDFS*; Addison-Wesley Professional: New York, NY, USA, 2016; pp. 4–18.

17. Ablimit, A.; Wang, F.; Vo, H.; Lee, R.; Liu, Q.; Zhang, X.; Joel, S. Hadoop GIS: A high performance spatial data warehousing system over mapreduce. *Proc. VLDB Endow.* **2013**, *6*, 1009–1020.

18. Ahmed, E.; Mohamed, F.M. Spatialhadoop: A mapreduce framework for spatial data. In Proceedings of the IEEE 31st International Conference on Data Engineering (ICDE), Seoul, Korea, 13–17 April 2015; pp. 1–12.

19. Louai, A.; Mohamed, F.M.; Mashaal, M. ST-HADOOP: A mapreduce framework for spatio-temporal data. *GeoInformatica* **2018**, *22*, 785–813.

20. Yu, J.; Zhang, Z.; Mohamed, S. Spatial data management in apache spark: The GeoSpark perspective and beyond. *GeoInformatica* **2018**. [CrossRef]

21. Brian, L. Geoprocessing in the Cloud. Available online: http://gsaw.org/wp-content/uploads/2014/10/2010s11d_levy.pdf (accessed on 20 August 2018).

22. Jason, L. GIS Tools for Hadoop: Big Data Spatial Analytics for the Hadoop Framework. Available online: http://esri.github.io/gis-tools-for-hadoop (accessed on 20 August 2018).

23. Ameet, K.; Rob, E. Geotrellis: Adding Geospatial Capabilities to Spark. Available online: https://databricks.com/session/geotrellis-adding-geospatial-capabilities-to-spark (accessed on 20 August 2018).

24. Li, Z.; Hu, F.; John, L.S.; Daniel, Q.D.; Tsengdar, L.; Michael, K.B.; Yang, C. A spatiotemporal indexing approach for efficient processing of big array-based climate data with MapReduce. *Int. J. Geogr. Inf. Sci.* **2017**, *31*, 17–35. [CrossRef]

25. Li, Z.; Huang, Q.; Gregory, J.C.; Hu, F. A high performance query analytical framework for supporting data-intensive climate studies. *Comput. Environ. Urban. Syst.* **2017**, *62*, 210–221. [CrossRef]

26. Gao, S.; Li, L.; Li, W.; Janowicz, K.; Zhang, Y. Constructing gazetteers from volunteered big geo-data based on Hadoop. *Comput. Environ. Urban. Syst.* **2017**, *61*, 172–186. [CrossRef]

27. Li, W.; Shao, H.; Wang, S.; Zhou, X.; Wu, S. A2CI: A cloud-based, service-oriented geospatial cyberinfrastructure to support atmospheric research. In *Cloud Computing in Ocean and Atmospheric Sciences*; Tiffany, C.V., Nazila, M., Yang, C., Yuan, M., Eds.; Elsevier: London, UK, 2016; pp. 137–161.

28. Li, Z.; Michael, E.H.; Li, W. A general-purpose framework for parallel processing of large-scale LiDAR data. *Int. J. Digit. Earth* **2018**, *11*, 26–47. [CrossRef]

29. Muhammad, U.R.; Anand, P.; Awais, A.; Chen, B.; Huang, B.; Ji, W. Real-time big data analytical architecture for remote sensing application. *IEEE J. Sel. Top. Appl. Earth Obs. Remote Sens.* **2015**, *8*, 4610–4621.

30. Sara, M.; Damiano, C.; Alberto, B. Adaptive Trip Recommendation System: Balancing Travelers among POIs with MapReduce. In Proceedings of the IEEE International Congress on Big Data, San Francisco, CA, USA, 2–7 July 2018; pp. 1–5.

31. Deepak, P.; Surya, N.; Rajiv, R.; Chen, J. A secure big data stream analytics framework for disaster management on the cloud. In Proceedings of the 18th IEEE International Conference on High Performance Computing and Communications, Sydney, Australia, 12–14 December 2016; pp. 1218–1225.

32. Addair, T.G.; Douglas, A.D.; Walter, W.R.; Stan, D.R. Large-scale seismic signal analysis with Hadoop. *Comput. Geosci.* **2014**, *66*, 145–154. [CrossRef]

33. Yellow River Institute of Hydraulic Research. *The Empirical Model of Annual Erosion and Sediment Production in Mesoscale Basins in Loess Plateau*; Yellow River Institute of Hydraulic Research: Zhengzhou, China, 2013; pp. 26–56.

34. Tom, W. *Hadoop: The Definitive Guide*, 4th ed.; O'Reilly Media: Sebastopol, CA, USA, 2015; pp. 19–41.

35. Mahmoud, P. *Data Algorithms: Recipes for Scaling Up with Hadoop and Spark*; O'Reilly Media: Sebastopol, CA, USA, 2015; pp. 39–58.

International Journal of
Geo-Information

Article

High-Performance Geospatial Big Data Processing System Based on MapReduce

Junghee Jo * and Kang-Woo Lee

IoT Research Division, Electronics and Telecommunications Research Institute (ETRI), 218 Gajeong-ro, Yuseong-gu, Daejeon 34129, Korea; kwlee@etri.re.kr
* Correspondence: dreamer@etri.re.kr

Received: 21 August 2018; Accepted: 4 October 2018; Published: 6 October 2018

Abstract: With the rapid development of Internet of Things (IoT) technologies, the increasing volume and diversity of sources of geospatial big data have created challenges in storing, managing, and processing data. In addition to the general characteristics of big data, the unique properties of spatial data make the handling of geospatial big data even more complicated. To facilitate users implementing geospatial big data applications in a MapReduce framework, several big data processing systems have extended the original Hadoop to support spatial properties. Most of those platforms, however, have included spatial functionalities by embedding them as a form of plug-in. Although offering a convenient way to add new features to an existing system, the plug-in has several limitations. In particular, while executing spatial and nonspatial operations by alternating between the existing system and the plug-in, additional read and write overheads have to be added to the workflow, significantly reducing performance efficiency. To address this issue, we have developed Marmot, a high-performance, geospatial big data processing system based on MapReduce. Marmot extends Hadoop at a low level to support seamless integration between spatial and nonspatial operations of a solid framework, allowing improved performance of geoprocessing workflow. This paper explains the overall architecture and data model of Marmot as well as the main algorithm for automatic construction of MapReduce jobs from a given spatial analysis task. To illustrate how Marmot transforms a sequence of operators for spatial analysis to map and reduce functions in a way to achieve better performance, this paper presents an example of spatial analysis retrieving the number of subway stations per city in Korea. This paper also experimentally demonstrates that Marmot generally outperforms SpatialHadoop, one of the top plug-in based spatial big data frameworks, particularly in dealing with complex and time-intensive queries involving spatial index.

Keywords: big data; IoT; MapReduce; Hadoop; geospatial big data; geospatial applications

1. Introduction

In the environment of the Internet of Things (IoT), various sensors have been mounted on objects in diverse domains, generating huge volumes of data at high speed [1,2]. A significant portion of sensor big data is geospatial data describing objects in relation to geographic information [3,4]. In general, geospatial big data refers to geographic data sets that cannot be processed using standard computing systems [3–5].

The general consensus among researchers from various domains is that "80% of data is geographic" [3,6–9]. The United Nations Initiative on Global Geospatial Information Management (UN-GGIM) reported that 2.5 quintillion bytes of data is created every day and a significant portion includes location components [10]. Google generates approximately 25 PB of data daily, a large portion of which consists of spatiotemporal characteristics [11]. The McKinsey Global Institute estimates the portion of spatial aspect data was about 1 PB in 2009 and is growing at an annual rate of 20% [12].

Due to the exponential growth of geospatial big data, utilization of data has become a global interest, increasingly drawing the attention not only of industry and academia, but also government agencies.

To promote openness and availability of geospatial big data, the US Federal Geographic Data Committee (FGDC) has defined the concept of National Spatial Data Infrastructure (NSDI) and preserves place-based data at all levels of government, private and nonprofit sectors, and academia. The FGDC provides rich sets of geospatial data to the public to support business, government agencies, or their partners. Geospatial One-Stop (GOS) is an initiative consistent with the goals of NSDI. It establishes a web-based portal (www.geodata.gov) for one-stop access to geospatial data and services and opens a developed portal to federal, state, and local governments, along with private citizens [13]. The EU directive, INSPIRE (Infrastructure for Spatial Information in Europe) project, plays an equivalent role for GOS [14,15]. It aims to benefit all levels of public authorities in Europe by producing and sharing integrated and quality geospatial information of EU Member States. In Oceania and Asia, there are also similar initiatives such as the Australian SDI [16] and Indian SDI [17], respectively.

The increasing volume and diverse sources (e.g., smart phones) of collected geospatial big data have created challenges in storing, managing, and processing data. In addition to the general characteristics of big data, the unique properties of spatial data such as computationally intensive processing of the time component make dealing with geospatial big data even more complicated. To some degree, traditional distributed and parallel processing frameworks make it possible to meet performance requirements for handling large-scale data. It has been several years since developers implemented an SQL-based system and operated it in a parallel way, though it was inadequate to deal efficiently with large volumes of data. System architecture has steadily evolved and a parallel database was designed to allow multiple data instances to share a single database with improvement in performance in parallel computing environments [18]. This is an efficient approach providing speedup and scale up for massive data, compared to traditional SQL systems.

Recently, several big data platforms have been built to facilitate developers of big data applications on a distributed and parallel computing platform. Particularly, Apache Hadoop [19] has existed for years and proven to be a mature and very popular platform for big data analysis for various applications. Hadoop is an open source MapReduce implementation being used at major corporations such as IBM, Amazon, Facebook, and Yahoo. Hadoop, however, is ill-equipped to support geospatial big data because its core structure does not consider the unique properties of spatial data (e.g., spatial data types). Also, the efficiency of operations (e.g., spatial queries) is limited on those platforms since the Hadoop's internal system is uninformed about spatial data.

To take advantage of the Hadoop/MapReduce environment in dealing with geospatial big data, several systems supporting spatial properties on Hadoop have emerged. ESRI has released a collection of GIS tools for spatial analysis [20] that provides access to the Hadoop system from ArcGIS products. In academia, Hadoop-GIS [21] is designed to support multiple types of spatial queries on MapReduce via skew-aware spatial partitioning to first partition spatial objects and then process them in parallel. Similarly, GPHadoop [22], a geoprocessing-enabled Hadoop platform, has been introduced to enable scalable geoprocessing to resolve geospatial related problems based on Hadoop and ESRI GIS tools. A main limitation of those Hadoop-based spatial systems is that they regard Hadoop as a black box and are therefore restricted by the limitations of the original Hadoop. To circumvent that issue, SpatialHadoop [23,24] has been designed, which injects spatial data awareness inside Hadoop making it more efficient to work with spatial data. SpatialHadoop, however, still presents a lack of integration with the core structure of Hadoop. For example, SpatialHadoop adds spatial data types or functions to the original Hadoop system as a form of plug-ins. However, compared to a solid framework, the performance of a plug-in based framework is limited by lacking seamless integration between spatial and nonspatial operations.

In this paper, we introduce the Marmot system, a Hadoop-based, high-performance data storage management system. It enables application developers having little working knowledge of big data technologies to implement high performance analysis tasks on geospatial big data. Compared with existing Hadoop-based spatial systems, the distinctive characteristics of Marmot can be summarized as follows.

- Marmot extends Hadoop at a low level and is a stream-based system supporting seamless integration between spatial and nonspatial operations in a solid framework.
- Marmot supports automatic construction of MapReduce jobs from a given spatial analysis task in ways to improve performance.
- Marmot allows developers to create various spatial applications by simply combining built-in spatial or nonspatial operators.

The rest of the paper is organized as follows. Section 2 shows our related work and Section 3 describes a systematic overview of Marmot including integration between spatial and nonspatial operations. Section 4 shows Marmot's data model and Section 5 presents the main algorithm which maps a sequence of operators for spatial analysis to MapReduce jobs and discusses implementation of Marmot. An example of a spatial application using Marmot and its performance evaluation is described in Sections 6 and 7, respectively. Finally, the discussion and conclusion along with our future work are presented in Sections 8 and 9.

2. Related Work

Several platforms have been developed for the processing of big data. MapReduce [25] is a distributed and parallel programming framework and is very popular due to its simplicity, scalability, and fault-tolerance. There are two phases in the MapReduce model: Map phase and Reduce phases. The Map phase is used to receive data and generate intermediate key and value pairs; the Reduce phase is used to accept the intermediate results and combine them into a single result. Hadoop [19] is an implementation of MapReduce as an open source form, which has been widely adopted in a majority companies as their technology stack.

Although Hadoop is known as the most representative distributed processing framework for handling big data, it cannot be used for real-time processing because the data are batched to each node for a given job. Spark [26] is an in-memory based high performance distributed data analysis system designed to address Hadoop's limitations. Unlike Hadoop, which is based on a batch processing engine, Spark can process data faster due to its in-memory computation policy. Although this feature provides advantages of low latency computation, there are also drawbacks such as requiring significantly more resources than Hadoop.

Most big data platforms, including Hadoop and Spark, have been designed without any specific consideration of spatial properties. A few platforms provide spatial functionalities [20–24], but they generally lack spatial properties and do not support advanced geospatial operations such as geospatial joins and geostatistical operations, which are imperative for advanced geospatial analytics. SpatialHadoop [22,23] has been developed to enable spatial operations by extending Hadoop layers: language, storage, MapReduce, and operations. Although SpatialHadoop has been known to overcome the limitations of existing spatial big data platforms and perform better than traditional Hadoop, it is still not sufficient in supporting in-depth geospatial analysis.

3. Marmot Overview

3.1. A Functional Architecture of High Performance

Marmot is a functional architecture that processes geospatial big data in parallel with MapReduce jobs for both batch and streaming services. A MapReduce job usually divides big data sets into separate chunks which are processed by map tasks in parallel. The framework sorts the results of the maps which are later input to reduce tasks.

Although MapReduce has been a highly popular framework for conducting parallelized distributed computing jobs, it is still not easy for developers to build complex geospatial big data applications. In fact, users of MapReduce often have analysis tasks which are too complicated to be transformed to MapReduce jobs. Such spatial tasks require handling the chain of multiple MapReduce steps where the output path of the data from previous MapReduce jobs becomes the input path to the next MapReduce job. To address this issue, Marmot supports automatic construction of one or more MapReduce tasks by taking on a spatial analysis task. When the task is transformed to MapReduce jobs, Marmot automatically controls the number of MapReduce jobs in a way to achieve better performance by decreasing the overhead of mapping and reducing.

Figure 1 shows a brief structure of Marmot which is based on a Hadoop framework. The core of Marmot consists of four main components: a plan scheduler, operation flow execution, geospatial big data storage management, and index management. This paper focuses on explaining how Marmot transforms a sequence of operators for spatial analysis to map and reduce functions. Specific content about other components is beyond the scope of this paper.

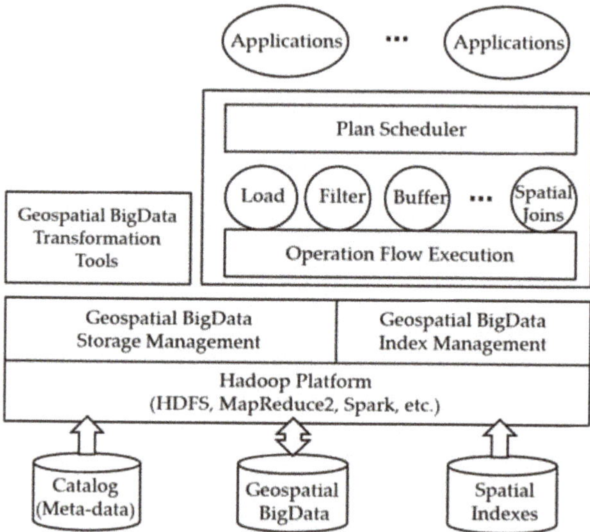

Figure 1. Overall structure of Marmot.

3.2. Integrating Spatial and Nonspatial Operations

NoSQL systems are very well-adapted to the environment offering high-performance information processing on a massive scale and so are increasingly used to deal with heavy demands of big data applications. Various NoSQL systems are currently available; however, only few provide spatial functionalities. MongoDB supports spatial indexing which allows users to execute spatial queries on a collection involving shapes and points, but it does not support many advanced geospatial operations which are essential for advanced geospatial analytics. PostGIS has more comprehensive geospatial functionalities but still provides only limited number of spatial data types and very few functions to handle them [27,28]. Thus, existing NoSQL systems needing to newly include or extend spatial functions generally use a form of plug-ins to support the extra feature within their systems. For example, Cassandra is extended to enable spatial data retrieval by extending its query language capabilities [28,29], and HBase is extended to support spatial indexes of Quad tree and K-d tree to enable range/kNN queries and point datasets [30].

Though the plug-in offers a convenient way to add new features to an existing system, it carries several limitations as described in Figure 2a. First, it is inadequate to support seamless interactions between spatial and nonspatial operations. A NoSQL system basically treats plug-ins as legacy and generates temporary files while executing spatial and nonspatial operations by turns, obstructing improvements in workflow. Second, its performance is limited due to the high number of input/output operations while reading or writing temporary files. Furthermore, this input/output overhead increases with increases in file size, as well as the number of operations. From this, we assume that if we reduce the data transfer overhead between NoSQL and spatial plugins, we could improve data processing performance.

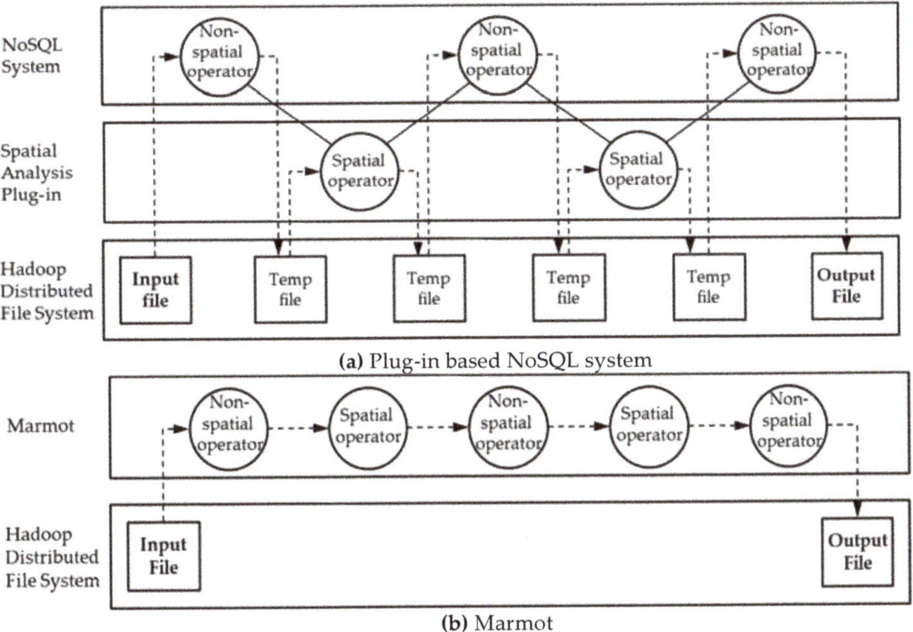

Figure 2. Comparison of two Hadoop-based architectures integrating spatial and nonspatial operations: (a) Plug-in based NoSQL system and (b) Marmot.

Marmot supports seamless interactions between spatial and nonspatial operations within a solid framework, as shown in Figure 2b. This is possible because Marmot is designed and developed by considering the data model and operations for processing spatial information from the initial phase of the development process. Most NoSQL systems, however, are designed with no consideration of spatial functionalities. This leads them to adopt plug-in based architecture when they want to embed spatial functionalities later on. Because both spatial and nonspatial operations are managed by one system, it is feasible to improve workflow. In addition, since temporary files do not need to be generated, the input/output overhead also decreases considerably regardless of file size or the number of operations for analysis.

4. Marmot Data Model

The term data model refers to an abstract model that arranges the structure of data and regulates their association with each other. Figure 3 shows the data model used in Marmot that corresponds mostly with the model in a standard relational database management system (RDBMS).

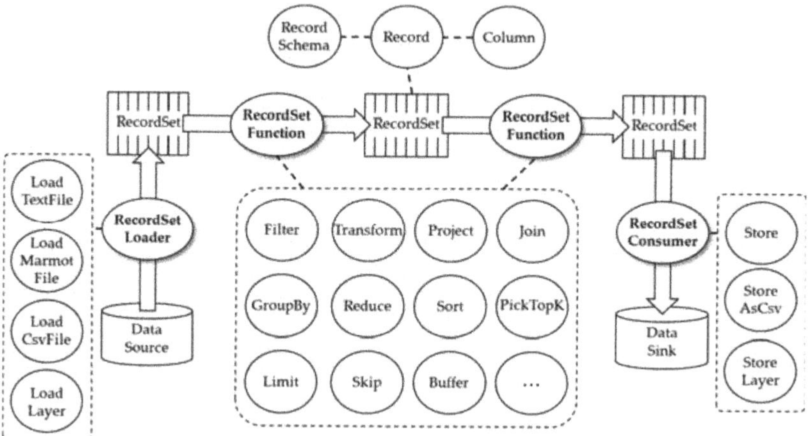

Figure 3. An overall description of a data model in Marmot.

In Marmot, a Record corresponds to a record in RDBMS, the smallest unit of a data element consisting of one or more column values within a table. The definition of a Record in Marmot is shown in Definition 1.

Definition 1. Record

```
public interface Record {
    public RecordSchema getSchema();

    public Object get(String colName);
    public Object get(int idx);
    public Object[[] getAll();

    public void set(String colName, Object value);
    public void set(int idx, Object value);
    public void set(Record rec, Boolean removeOverflow);

    public Geometry getGeometry( ... ... );
    public String getString( ... ... );
}
```

A RecordSet is a set of records corresponding to a table in RDBMS which provides forward-only access to a data source. A RecordSchema represents metadata consisting of information containing the names and types of columns composing a Record. All the Records included in a specific RecordSet are compliant with the same RecordSchema. The definitions of RecordSet and RecordSchema are shown in Definition 2 and Definition 3, respectively.

Definition 2. RecordSet

```
public interface RecordSet {
    RecordSchema getRecordSchema();
    Boolean next(Record record);
    Record nextCopy();
    void close();

    Stream<Record> stream();
    List<Record> toList();
    Iterator<Record> iterator();
}
```

Definition 3. RecordSchema

```
public class RecordSchema {
    public int getColumnCount();

    public Boolean existsColumn(String colName);
    public Column getColumn(String colName);
    public Column getColumnAt(int idx);
    public Collection<Column> getColumnAll();
}
```

A RecordSetOperator is a processing unit using a RecordSet which corresponds to the concept of a relational operator in RDBMS. As the input and output of relational operators in RDBMS are a table, the input and output of a RecordSetOperator is a RecordSet. In Marmot, there are three types of a RecordSetOperator, as shown in Figure 4.

Figure 4. A description of a RecordSetOperator in Marmot: RecordSetLoader, RecordSetFuntion, and RecordSetConsumer.

A RecordSetLoader is an operator that loads input data from outside data sources and converts it to RecordSet to be handled within Marmot. For example, loadTextFile reads a text file from a given path and converts it to a RecordSet. A RecordSetConsumer stores a RecordSet created within Marmot as the final result of a given analysis to a specified outside path. A Marmot code describing RecordSetLoader and RecordSetConsumer is given in Figure 5.

```
Plan plan = new PlanBuilder()

.load("/traffic/subway/stations")      ⎤ RecordSetLoader

.filter("transit_yn = 'y')

.project("the_geom,station_id")

.store("output/transit_stations");     ⎤ RecordSetConsumer
```

Figure 5. A Marmot code of an example of RecordSetLoader and RecordSetConsumer.

Lastly, the RecordSetFunction is an operator that takes a RecordSet as input data and outputs a new RecordSet. In Marmot, there are two types of a RecordSetFuntion: spatial operators and nonspatial operators. Spatial operators are functions taking input spatial data, analyzing various types of spatial relationships, and producing output spatial data. For example, Buffer is a spatial operator creating a new RecordSet which is the result of encompassing the area around a specific location based on a predefined distance. Project and Filter are examples of nonspatial operators, where Project produces a RecordSet composed only of specified columns; Filter generates a RecordSet consisting only of columns satisfying certain conditions.

Table 1. Spatial and nonspatial operators in RecordSetFuntion supporting in Marmot.

Spatial operators	Nonspatial operators
Buffer, Centroid, TransformCRS, Dissolve, LookupPostalAddress, AggrUnion, AggrConvexHull, Intersection, SpatialJoin, SpatialOuterJoin, SpatialSemiJoin, kNNJoin, ClipJoin, DifferenceJoin, AggregateJoin, LoadSpatialIndexJoin, AggrEnvelope, RangeQuery, EstimateIDW, GetisOrd, EstimateKernelDensity, LocalMoran's I, E2SFCA	Filter, Project, Update, Transform, Skip, Distinct, Rank, PickTopK, Aggregate, GroupBy, Shard, Sample, Sort, Limit, Reduce, MapReduceEquiJoin, ReduceByGroupKey

The spatial and nonspatial operators currently supported in Marmot are shown in Table 1. Examples of their interfaces and usages are shown in Examples 1 and 2.

Example 1. Nonspatial operators in RecordSetFuntion

- filter(predicate_str)
  [Usage] filter("age > 20")

- project(column_comma_list)
  [Usage] project("the_geom,id,age")

- transform(File scriptFile, dropInputCols)
  [Usage] transform(new File("transform_script.xml"), true)

- pickTopK(cmpColSpecs, topK)
  [Usage] pickTopK("name,age:D", 10)

- groupBy(key_cols_comma_list)
  [Usage] groupBy("car_no, driver_id")

Example 2. Spatial operators in RecordSetFuntion

- loadSpatialIndexJoin(condition, layer1, layer2, result_column1, result_colume2)
  [Usage] loadSpatialIndexJoin(SpatialPredicate.INTERSECTS,
          "admin/cadastral/clusters", "admin/urban_area/clusters", "*",
          "*-{the_geom},the_geom as the_geom2")

---

5. Spatial Analysis using Marmot

5.1. Plan: A Chain of RecordSetOperators

Spatial analysis is the process of investigating the positions, attributes, or relationships of objects based on spatial data to obtain valuable knowledge or to answer a question. In Marmot, spatial analysis using big data is defined as a chain of one or more units of RecordSetOperators, called a Plan. In a Plan, a RecordSetLoader and a RecordSetConsumer are generally placed in the first and last position, respectively; each RecordSetFunction is placed in the middle of the Plan. In order to process a given Plan, Marmot sequentially processes each RecordSetOperator comprising the Plan.

Figure 6 shows an example of a Plan consisting of a total of five RecordSetOperators: one RecordSetLoader, three RecordSetFunctions, and one RecordSetConsumer. As mentioned in the previous section, the RecordSetLoader transforms external geospatial big data into an internal form of Marmot which is a RecordSet. Then, one or more numbers of RecordSetFuntions process the RecordSet and create a new RecordSet containing the desired contents. Finally, the RecordSetConsumer stores the result in the output files on the file system.

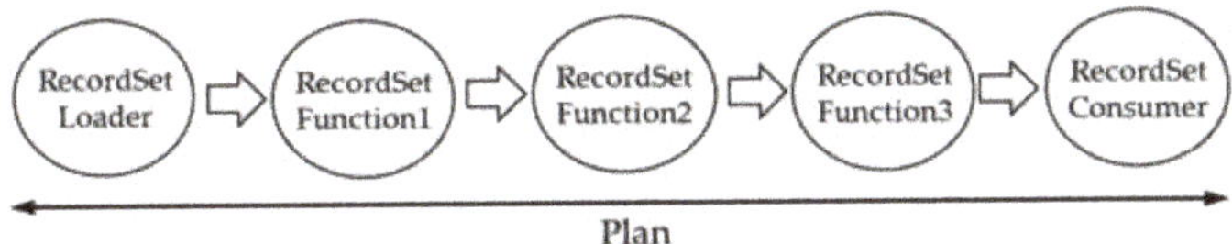

Figure 6. A description of a Plan in Marmot: A chain of RecordSetOperators.

5.2. Mapping a Plan to a Sequence of MapReduce Jobs

Marmot receives a Plan and automatically constructs one or more MapReduce jobs based on the features of each of the RecordSetFunctions. In general, to create a MapReduce job for a required operation in Hadoop, a programmer needs to specify both a map function and a reduce function. A spatial analysis task usually requires multiple MapReduce jobs to achieve a result, thereby making it cumbersome to write iterative code to describe complex MapReduce workflows [31]. The main idea of Marmot is to provide advantages which enable developers that have no detailed knowledge of MapReduce to build high performance geospatial analysis applications by simply combining built-in RecordSetFuntions. The specific process of mapping a Plan to a sequence of MapReduce jobs is described in Figure 7.

In Marmot, operators in RecordSetFuntions are classified into one of three categories: MapReduceTerminal, MapReduceJoint, or neither, depending on their own features. The term MapReduceTerminal implies operators not generating a RecordSet as the result of their executions; operators involved in RecordSetConsumer (e.g., storeAsCsv) usually belong to this category. The term MapReduceJoint is defined as those operators in RecordSetFuntions generating a RecordSet after grouping input data based on specific columns and then processing the grouped data. Aggregation operators (e.g., GroupBy) performing an operation on a given set of values and returning a RecordSet usually belong to this category. Lastly, there are operators in RecordSetFuntions not involved in either MapReduceTerminal or MapReduceJoint—most operators belong to this category. Among the three categories, MapReduceJoint is an important separation point to produce the Map phase and Reduce phase. An example of MapReduceTerminal and MapReduceJoint is shown in Table 2.

The process of mapping a Plan to a sequence of MapReduce jobs, specified in Figure 7, is as follows. When Marmot receives a Plan from an analysis application, Marmot extracts the first RecordSetOperator from the Plan and checks whether the operator belongs to the category of MapReduceJoint. If the operator does not belong to MapReduceJoint, Marmot inserts it in a separate list storing RecordSetOperators to be executed in Map phase (i.e., M in Figure 7) and extracts the next RecordSetOperator. This process is repeated until the last RecordSetOperator is inserted into the M then, finally, a MapReduce job is created consisting of only one Map phase.

During the process of constructing M when Marmot finds an operator belonging to MapReduceJoint, it constructs another list which is a set of RecordSetOperators to be executed in Reduce phase (i.e., R in Figure 7). The specific procedure of constructing R is as follows. Marmot decomposes the operator belonging to MapReduceJoint and extracts two suboperators: the mapping operator and reducing operator. For example, in the case of PickTopK which is one of the RecordSetOperators in MapReduceJoint, Sort (cols) and Take (k) are mapping and reducing operators of PickTopK, respectively. The extracted mapping operator is attached at the end of M and the reducing operator is inserted into the initialized R.

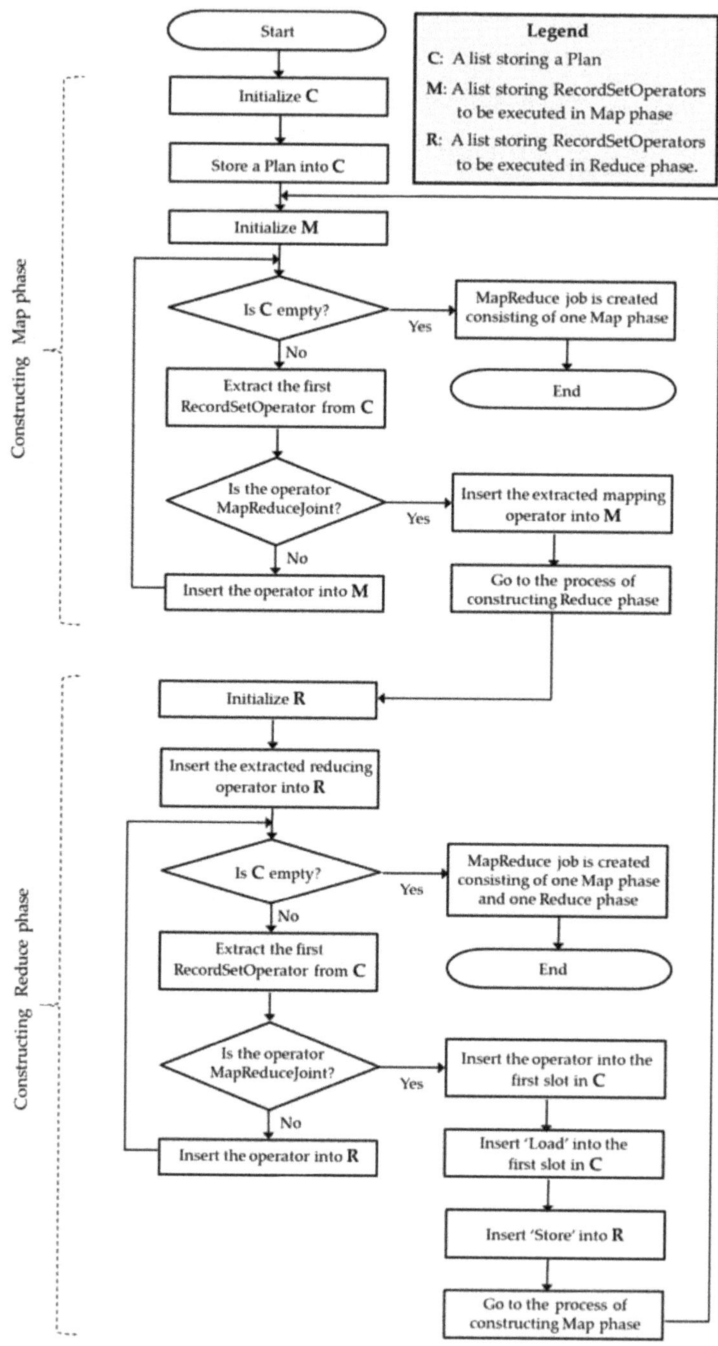

Figure 7. An algorithm of automatic mapping a Plan to a sequence of MapReduce jobs in Marmot. C refers to a list storing a Plan; M refers to a list storing RecordSetOperators to be executed in Map phase; and R refers to a list storing RecordSetOperators to be executed in Reduce phase.

Marmot then extracts the next RecordSetOperator from the Plan and resumes examining whether the operator belongs to the category of MapReduceJoint. If the operator belongs to MapReduceJoint, Marmot inserts it in R and extracts the next RecordSetOperator. This process is repeated until the last RecordSetOperator is inserted in the R; then finally a MapReduce job is created consisting of one Map phase and one Reduce phase. During the process of constructing *R*, when Marmot meets an operator belonging to MapReduceJoint, Marmot iteratively constructs another MapReduce job by treating the remaining RecordSetOperators as a new Plan.

Table 2. RecordSetOperators classified as MapReduceTermial and MapReduceJoint, repectively.

MapReduceTerminal	MapReduceJoint
storeMarmotFile, store, storeAsCsv	MapReduceEquiJoin, PickTopK, ReduceByGroupKey

Figure 8 shows an example of the process of creating MapReduce jobs in Marmot. The given Plan in the example consists of five RecordSetOperators where operator3 is the only operator belonging to MapReduceJoint. During the process, operator3 is decomposed into two operators—a mapping operator and reducing operator. The Plan is eventually transformed into MapReduce jobs consisting of several map tasks and reduce tasks. Each map task performs operator1 and operator2 and the reduce tasks perform operator4 and operator5. Operator3, which is defined as a MapReduceJoint, is broken down into two suboperators and they are carried out at map tasks and reduce tasks, respectively. The number of map tasks is heavily dependent on the size of the input dataset. In the default configuration, each map task takes care of 64MB of input dataset. The number of reduce tasks is, however, is usually decided by the parameters of the MapReduceJoint operator, which is operator3 in this example.

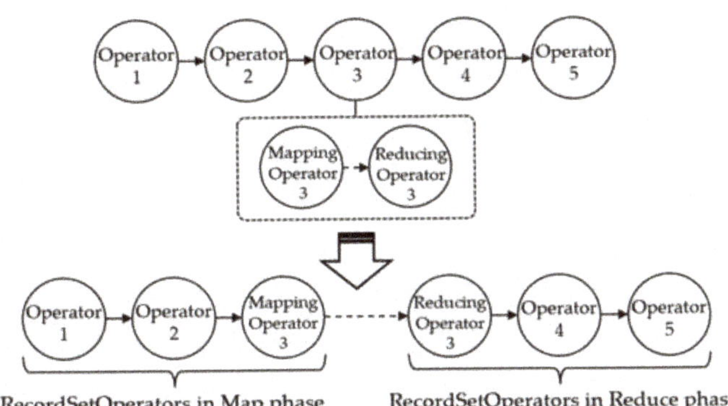

Figure 8. An example of transforming a Plan to a sequence of MapReduce jobs in Marmot.

Although Marmot has been built on a CentOS 6.9 environment, it can easily be executed on any operating system where Hadoop is installed due to the characteristics of Java's platform independency. As a Hadoop system, Marmot uses Hortonworks Data Platform (HDP) 2.6 version. Table 3 shows the specific environment of implementation and testing of Marmot.

Table 3. Implementation and testing environment for Marmot.

Software	Version
CentOS	6.7
Hortonworks Data Platform	2.6
JDK	1.8
PostgreSQL	9.4.7

6. Case Analysis: Subway Stations

To show how Marmot handles geospatial big data, this section presents an example of spatial analysis. Figure 9 is a sample code of a Plan retrieving the number of subway stations per city. For this analysis, nationwide datasets of cadastral and subway stations in Korea are used.

```
Plan plan;

plan = marmot.planBuilder("Top five cities with the largest number of subway stations")
            .load("subway_stations")
            .centroid("the_geom")
            .spatialJoin("the_geom", "cadastral", "the_geom")
            .groupBy("city_id")
                .aggregate(COUNT())
            .pickTopK("count:DESC", 5)
            .storeAsCsv("result")
            .build();
```

Figure 9. A Marmot code of an example Plan retrieving the top five cities with the largest number of subway stations.

The plan is composed of six RecordSetOperators: 'Load', 'Centroid', 'SpatialJoin', 'GroupBy', 'PickTopK', and 'StoreAsCsv'. As shown in Figure 10, using the 'Load' operator, Marmot reads the boundaries of each subway station and 'Centroid' operator computes their center coordinates. With the 'SpatialJoin' operator, Marmot finds a city whose boundary contains the centroid of each subway station. Then, the 'GroupBy' operator groups the output records based on each city and counts the number of their records. The output records should be '(city_id, count)' pairs, which contain the number of subway stations for each city. Finally, the 'PickTopK' operator sorts the record on the basis of numbers and picks the top five cities. These five records are stored in the Hadoop Distributed File System (HDFS) by the 'StoreAsCsv' operator.

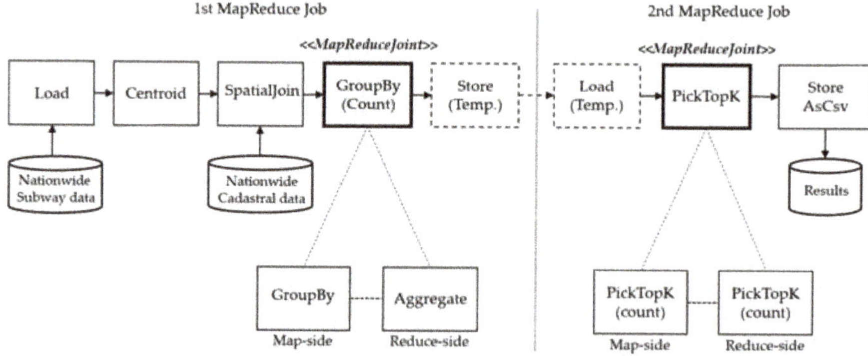

Figure 10. A sequence of RecordSetOperators consisting of the example Plan.

In the process of transforming the Plan to a sequence of MapReduce jobs, Marmot recognizes two 'MapReduceJoint' operators: 'GroupBy' and 'PickTopK'. Since each 'MapReduceJoint' generates one MapReduce job, the plan is transformed into two separate MapReduce jobs, as depicted in Figure 10. For the first MapReduce job, the 'GroupBy' operator is the breakpoint producing the first Map and

Reduce tasks, and 'PickTopK' operator is another breakpoint to produce the second MapReduce job. The output of the first MapReduce job is stored in a temporary HDFS file—so that the second MapReduce job can continue to execute the rest of the given Plan. Marmot deletes the temporary file when the whole plan has been executed.

The sequence of RecordSetOperators assigned during Map and Reduce phases is executed in a pipeline manner by a number of mappers and reducers. The number of mappers is dynamically determined by the size of input data whereas the number of reducers is statically fixed in a Plan. During the process of a given analysis, four mappers with two reducers are used in the first MapReduce job and two mappers with two reducers are used in the second MapReduce job, as shown in Figure 11.

1st MapReduce Job

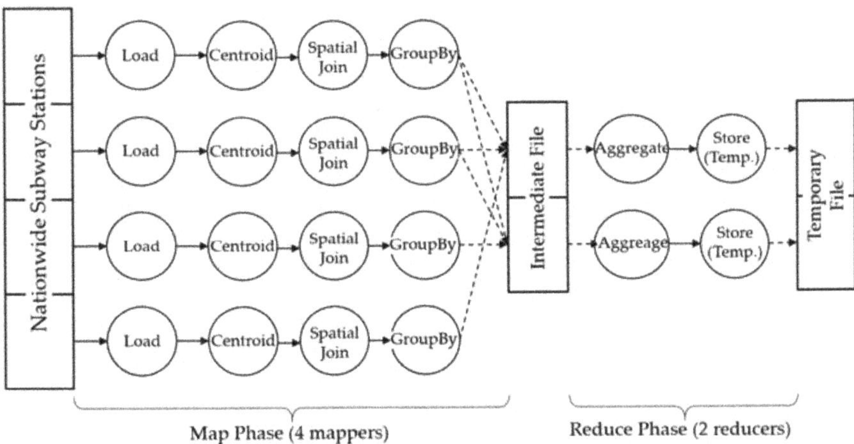

2nd MapReduce Job

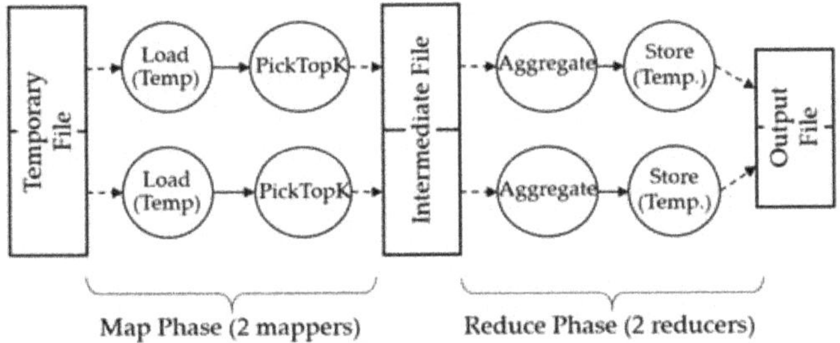

Figure 11. Operations in Map and Reduce phases in the first and second MapReduce jobs.

7. Experiments

7.1. Test Cases

Experiments were conducted to compare the performance of Marmot to that of SpatialHadoop, one of the top MapReduce frameworks supporting spatial functionalities. Since the geospatial operators currently supported by SpatialHadoop are limited compared to Marmot, it is difficult to conduct our experiments based on complex spatial analysis tasks. Therefore, we decided to perform a comparison for each of the fundamental spatial operations supported by both SpatialHadoop and Marmot.

There are five target spatial operations in our experiments: obtaining minimum bounding rectangle (MBR) and creating spatial index, range query without index, range query with index, and spatial join with index. MBR has been most commonly used to approximate spatial objects and is one of the fundamental operations for spatial analysis. A query for obtaining MBRs is therefore included in our experiment. Spatial index also plays an important role in geospatial domain to speed up retrieving certain objects in a spatial database, making measuring the performance of creating spatial index essential for this evaluation. Range query allows one to search for spatial objects located in a specified spatial extent and is therefore a fundamental type of query in spatial databases. To see the impact of using spatial index on range query performance, we investigated two cases—with and without an index. Lastly, spatial join is one of the most important operations for combining spatial objects and serves as building blocks for processing complex spatial analysis. Even with the support of spatial index, spatial join is particularly complex and time-intensive. Thus, efficient processing of spatial join is crucial to increase query performance.

7.2. Data Description

For the experiment, we used two Tiger Files including real spatial data for Korea. One file contained a nationwide continuous cadastral map with a size of 16GB (38,744,510 polygons); the other file included major urban areas created for management of land use with a size of 0.3GB (53,087 polygons). Figure 12 shows the visualization of these two Tiger Files including real spatial datasets of Korea. Particularly, Figure 12a shows just part of the continuous cadastral map because the entire data are too large to load onto the visualization tool. To demonstrate what all the data look like, partial data with a size of 1.24GB (4,720,616 polygons) involving Seoul city and Gyeonggi Province are plotted using red on a grayscale OSM basemap as the one by Stamen.

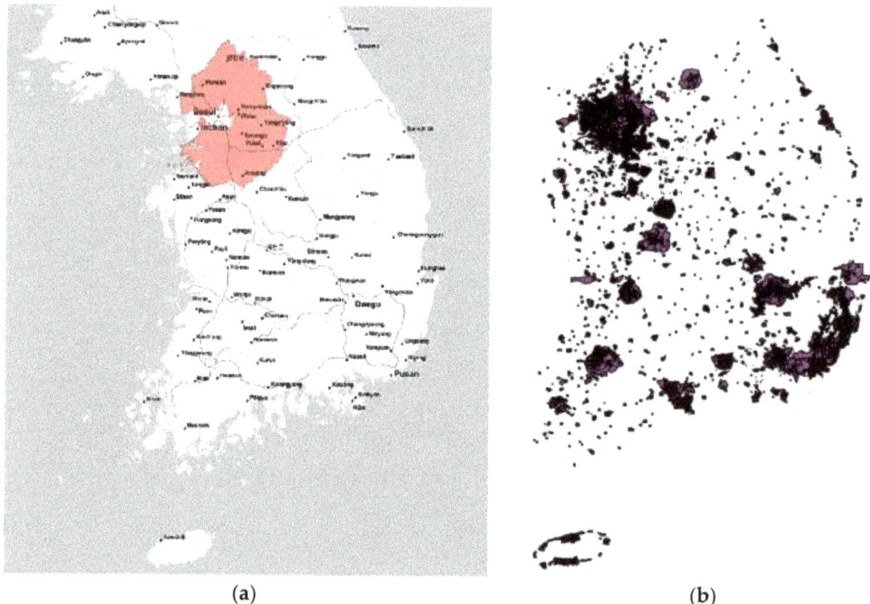

(a)   (b)

Figure 12. Visualization of two Tiger Files including real spatial datasets of Korea. (**a**) Part of the continuous cadastral map (1.24GB out of 16GB; 4,720,616 polygons out of 38,744,510 polygons) plotted using red and (**b**) major urban areas (0.3GB, 53,087 polygons) created for management of land use.

Of the five test cases in our experiment, three cases (i.e., obtaining MBR, range query without index, and range query with index) use only the cadastral data; one case (i.e., creating spatial index) uses only the major urban data and one case (i.e., spatial join) uses both the cadastral and major urban data.

7.3. Results

Each test query was run five times and the average, after excluding the highest and lowest values, was chosen to evaluate performance. The formula to express the performance improvement rate (PIR) of Marmot is:

$$PIR = ((SH - M)/M) * 100 \qquad (1)$$

where SH and M are execution times required by SpatialHadoop and Marmot, respectively. Table 4 shows the execution time of each test case and the PIR.

Table 4. A comparison of execution time and performance improvement rate (PIR) (in seconds).

Geospatial operators	SpatialHadoop	Marmot	PIR
Create MBR	81.634	20.336	301%
Create spatial index	220.494	95.220	132%
Spatial join with index	21.841	12.688	72%
Range query with index	2.836	0.918	209%
Range query without index	17.827	16.972	5%

In designing the experiment, we included only fundamental spatial operations due to the infrastructural limits of SpatialHadoop. Although SpatialHadoop has built-in spatial analytic functions, it supports few spatial operations and lacks many useful functions such as coordinate conversation, exporting spatial data, and raster processing functions. In addition, SpatialHadoop was not designed to read Shapefiles directly, which is a very popular geospatial vector data format used in spatial domain. Instead, SpatialHadoop reads spatial data represented in well-known text (WKT) or well-known binary (WKB). Because of this, in this paper we focused on measuring only the performance of fundamental spatial operations.

In all five test cases, Marmot outperforms SpatialHadoop with a variation depending on the type of test. In the case of simple operations such as range query, the performance difference between SpatialHadoop and Marmot is small (PIR = 5% for range query without index). However, in the case of complex operations especially involving spatial index, Marmot highly outperforms SpatialHadoop (PIR = 209% for range query with index; PIR = 72% for spatial join with index). This is because those operations are strongly influenced by the performance of spatial index and Marmot greatly outperforms SpatialHadoop in these operations (PIR = 132% for creating spatial index). Lastly, in the case of creating MBR that is an essential prerequisite for creating spatial index, Marmot also has a higher performance than SpatialHadoop (PIR = 301% for creating MBR).

8. Discussion

Marmot has been developed as one component for constructing a national geospatial big data platform [4] promoted by the Ministry of Land, Infrastructure, and Transport of the Korean government. Marmot is publicly available including GitHub [32] and will be integrated into other components of the platform to be used for public services such as transport, real estate, or disaster prevention. Based on the long-term plans of the Korean government to promote the geospatial industry, Marmot also will be integrated with the national geospatial information open platform, V-World [33], to provide a variety of map-based spatial analysis services. In addition, other ministries dealing with spatial data also plan to use Marmot as an infrastructure platform to provide big data based analysis services.

Compared to existing systems, Marmot is distinguished by the following. First, Marmot supports seamless integration between spatial and nonspatial operations within a solid framework, thereby improving workflow performance. Second, Marmot outperforms existing top-tier, plug-in based, spatial big data frameworks, especially for complex and time-intensive queries involving a spatial index. Third, Marmot provides a variety of spatial operators, which allow executing further complex spatial analyses. Finally, once application developers recognize a set of nonspatial and spatial operators, they can implement desired spatial analysis tasks without having to possess detailed knowledge of big data technologies.

Although Marmot shows good performance results compared to other existing research, it still needs further improvement. First, Marmot is restricted to batch processing only. Batch processing is efficient for handling huge amounts of data, but outcomes can be delayed depending on the input data size and computing power of a machine. Many recent geospatial data analytics require real-time data processing; however, Marmot does not yet support real-time processing of streamed data. Second, many machine learning algorithms read modestly sized source datasets iteratively, e.g., K-means and DBSCAN. For those applications, in-memory data processing offers the best match, but the current version of Marmot lacks this. Although Marmot utilizes memory, in part, along with disks during data processing [34], it cannot take full advantage of in-memory computation to increase processing speed.

In the future, we first plan to investigate what additional factors create performance differences between Marmot and SpatialHadoop and conduct extensive experiments using more large-scale geospatial big data. Second, we are going to finalize implementing an Apache Spark-based geospatial big data processing system, which is currently underdeveloped. To address the aforementioned limitations of Marmot, we have selected Spark as a framework for our next system for analyzing geospatial big data since Spark supports both in-memory and real-time data processing. Finally, we will design and conduct experiments comparing performance between Marmot and the new system.

9. Conclusions

The majority of existing NoSQL systems supporting geospatial data processing on Hadoop embed spatial functions as a form of plug-in to support extra features within their systems. This plug-in based approach suffers from severe performance problems due to the massive volume of data transfer between NoSQL and spatial plugins. In order to reduce data-transfer overhead, we have developed a new geospatial data processing framework on Hadoop, named Marmot. Marmot executes both spatial operations and nonspatial operations in the same session to avoid massive data transfer between operations, and employs a new method that maps a sequence of operations, either spatial or nonspatial, into a series of MapReduce jobs. We also show the proposed method outperforms SpatialHadoop with performance experiments.

Author Contributions: Kang-Woo Lee designed and implemented Marmot; Kang-Woo Lee and Junghee Jo conducted the testing of Marmot and analyzed the results; Junghee Jo wrote the manuscript; and Kang-Woo Lee revised the manuscript.

Acknowledgments: This research, "Geospatial Big Data Management, Analysis, and Service Platform Technology Development", was supported by the MOLIT (The Ministry of Land, Infrastructure, and Transport), Korea, under the National Spatial Information Research Program supervised by the KAIA (Korea Agency for Infrastructure Technology Advancement) (18NSIP-B081011-05).

Conflicts of Interest: The authors declare no conflicts of interest.

References

1. Yang, C.; Liu, C.; Zhang, X.; Nepal, S.; Chen, J. A time efficient approach for detecting errors in big sensor data on cloud. *IEEE Trans. Parallel Distrib. Syst.* **2015**, *26*, 329–339. [CrossRef]
2. Ang, L.M.; Seng, K.P.; Zungeru, A.; Ijemaru, G. Big sensor data systems for smart cities. *IEEE Internet Things J.* **2017**, *4*, 1259–1271. [CrossRef]
3. Li, S.; Dragicevic, S.; Castro, F.A.; Sester, M.; Winter, S.; Coltekin, A.; Pettit, C.; Jiang, B.; Haworth, J.; Stein, A.; et al. Geospatial big data handling theory and methods: A review and research challenges. *ISPRS J. Photogramm. Remote Sens.* **2016**, *115*, 119–133. [CrossRef]

4. Lee, J.G.; Kang, M. Geospatial big data: Challenges and opportunities. *Big Data Res.* **2015**, *2*, 74–81. [CrossRef]
5. Amirian, P.; Van Loggerenberg, F.; Lang, T.; Varga, M. Geospatial big data for finding useful insights from machine data. In *GISResearch UK*; University of Leeds: Leeds, United Kingdom, 2015.
6. Morais, C.D. Where is the Phrase "80% of Data is Geographic" From? Available online: http://www.gislounge.com/80-percent-data-is-geographic (accessed on 4 April 2018).
7. Jeansoulin, R. Review of forty years of technological changes in geomatics toward the big data paradigm. *ISPRS Int. J. Geo-Inf.* **2016**, *5*, 155. [CrossRef]
8. He, Z.; Liu, Q.; Deng, M.; Xu, F. March. Handling multiple testing in local statistics of spatial association by controlling the false discovery rate: A comparative analysis. In Proceedings of the 2017 IEEE 2nd International Conference Big Data Analysis (ICBDA), Beijing, China, 10–12 March 2017; pp. 684–687.
9. Seethapathy, B.K.; Parvathi, R. A review on spatial big data analytics and visualization. In *Modern Technologies for Big Data Classification and Clustering*; IGI Global: Philadelphia, PA, USA, 2017; p. 179. ISBN 978-1522528050.
10. Carpenter, J.; Snell, J. Future Trends in Geospatial Information Management: The Five to Ten Year Vision. United Nations Initiative on Global Geospatial Information Management. Available online: http://ggim.un.org/ggim_20171012/docs/meetings/GGIM5/Future%20Trends%20in%20Geospatial%20Information%20Management%20%20the%20five%20to%20ten%20year%20vision.pdf (accessed on 4 April 2018).
11. Vatsavai, R.R.; Ganguly, A.; Chandola, V.; Stefanidis, A.; Klasky, S.; Shekhar, S. Spatiotemporal data mining in the era of big spatial data: Algorithms and applications. In Proceedings of the 1st ACM SIGSPATIAL International Workshop on Analytics for Big Geospatial Data, Redondo Beach, CA, USA, 7–9 November 2012; pp. 1–10.
12. Dasgupta, A. Big Data: The Future is in Analytics. Available online: https://www.geospatialworld.net/article/big-data-the-future-is-in-analytics (accessed on 20 April 2018).
13. Maguire, D.J.; Longley, P.A. The emergence of geoportals and their role in spatial data infrastructures. *Comput. Environ. Urban Syst.* **2005**, *29*, 3–14. [CrossRef]
14. Annoni, A.; Bernard, L.; Fullerton, K.; de Groof, H.; Kanellopoulos, I.; Millot, M.; Peedell, S.; Rase, D.; Smits, P.; Vanderhaegen, M. Towards a European spatial data infrastructure: The INSPIRE initiative. In Proceedings of the 7th International Global Spatial Data Infrastructure Conference, Bangalore, India, 2–6 February 2004.
15. Bernard, L.; Kanellopoulos, I.; Annoni, A.; Smits, P. The European geoportal—One step towards the establishment of a European Spatial Data Infrastructure. *Comput. Environ. Urban Syst.* **2005**, *29*, 15–31. [CrossRef]
16. Busby, J.R.; Kelly, P. Australian spatial data infrastructures. In Proceedings of the 7th International Global Spatial Data Infrastructure Conference, Bangalore, India, 2–6 February 2004.
17. Sivakumar, R.; Rao, M.; Dasgupta, A.R. Perspectives of India's national spatial data infrastructure. In Proceedings of the 7th International Global Spatial Data Infrastructure Conference, Bangalore, India, 2–6 February 2004.
18. Dewitt, D.; Gray, J. Parallel database system: The future of high performance database systems. *Commun. ACM* **1992**, *35*, 85–98. [CrossRef]
19. White, T. *Hadoop: The Definitive Guide*, 3rd ed.; O'Reilly Media: Newton, MA, USA, 2012; ISBN 1449338771.
20. GitHub. Available online: http://esri.github.io/gis-tools-for-hadoop (accessed on 10 May 2018).
21. Aji, A.; Wang, F.; Vo, H.; Lee, R.; Liu, Q.; Zhang, X.; Saltz, J. Hadoop-GIS: A high performance spatial data warehousing system over MapReduce. In Proceedings of the VLDB Endowment 2013, Trento, Italy, 26–30 August 2013; pp. 1009–1020.
22. Gao, S.; Li, L.; Li, W.; Janowicz, K.; Zhang, Y. Constructing gazetteers from volunteered big geo-data based on Hadoop. *Comput. Environ. Urban Syst.* **2017**, *61*, 172–186. [CrossRef]
23. Eldawy, A. SpatialHadoop: Towards flexible and scalable spatial processing using mapreduce. In Proceedings of the SIGMOD PhD Symposium 2014, Snowbird, UT, USA, 22–27 June 2014; ACM: New York, NY, USA; pp. 46–50.
24. Eldawy, A.; Mokbel, M.F. SpatialHadoop: A MapReduce framework for spatial data. In Proceedings of the 2015 IEEE 31st International Conference Data Engineering (ICDE), Seoul, South Korea, 13–17 April 2015; pp. 1352–1363.
25. Dean, J.; Ghemawat, S. MapReduce: Simplified data processing on large clusters. *Commun. ACM* **2008**, *51*, 107–113. [CrossRef]

ISPRS Int. J. Geo-Inf. **2018**, *7*, 399

26. Zaharia, M.; Chowdhury, M.; Franklin, M.J.; Shenker, S.; Stoica, I. Spark: Cluster computing with working sets. In Proceedings of the 2nd USENIX Conference on Hot Topics in Cloud Computing, Boston, MA, USA, 21–25 June 2010.
27. Agarwal, S.; Rajan, K.S. Analyzing the performance of NoSQL vs. SQL databases for Spatial and Aggregate queries. In Proceedings of the Free and Open Source Software for Geospatial (FOSS4G) Conference 2017, Boston, MA, USA, 14–19 August 2017.
28. Brahim, M.B.; Drira, W.; Filali, F.; Hamdi, N. Spatial data extension for Cassandra NoSQL database. *J. Big Data* **2016**, *3*, 11. [CrossRef]
29. Vasavi, S.; Priya, M.P.; Gokhale, A.A. Framework for Geospatial Query Processing by Integrating Cassandra with Hadoop. *Knowl. Comput. Appl.* **2018**, 131–160. [CrossRef]
30. Eldawy, A.; Mokbel, M.F. The ecosystem of SpatialHadoop. *SIGSPATIAL Spec.* **2015**, *6*, 3–10. [CrossRef]
31. Manoochehri, M. *Data Just Right: Introduction to Large-Scale Data & Analytics*; Addison-Wesley: Boston, MA, USA, 2013; ISBN 978-0133359077.
32. Marmot from GitHub. Available online: https://github.com/kwlee0220/marmot.server.dist (accessed on 29 September 2018).
33. Ministry of Land, Infrastructure and Transport in Korea, V-World. Available online: http://eng.vworld.kr/eng/em_main.do (accessed on 15 September 2018).
34. Jo, J.H.; Lee, K.W. Marmot: A Hadoop-based high performance data storage management system for processing geospatial or geo-spatial big data. *Korean Soc. Geospat. Inf. Sci.* **2018**, *26*, 3–10.

MDPI
St. Alban-Anlage 66
4052 Basel
Switzerland
Tel. +41 61 683 77 34
Fax +41 61 302 89 18
www.mdpi.com

ISPRS International Journal of Geo-Information Editorial Office
E-mail: ijgi@mdpi.com
www.mdpi.com/journal/ijgi